General Educational Development Testing Service
A Program of the American Council on Education®

Dear GED Candidate:

Congratulations on taking one of the most important steps of your life—earning a GED credential!

Since 1942, millions of people like you have taken the GED Tests to continue their education, to get a better job, or to achieve a sense of accomplishment.

We are delighted to introduce **Keys to GED® Success: Science**—an invaluable resource to help you pass the GED Science Test. It has been developed through a partnership between the GED Testing Service®–developer of the GED Tests–and Steck-Vaughn, a leading provider of GED test preparation materials and the exclusive distributor of the Official GED Practice Tests.

GEDTS researched the types of skills that GED candidates could focus on to improve their chances of passing the tests. We identified the types of questions and possible reasons that test-takers were missing specific questions on each test and decided to share that information. GEDTS collaborated with Steck-Vaughn to target those skills in a workbook that would benefit present and future GED candidates. The skills targeted in our research are called the **GED® Key Skills**—which is what you'll find in this book. In addition to the **GED Key Skills**, this book includes other important lessons that are needed to pass the GED Science Test.

To help GED teachers, there is a Teaching Tips section included. The tips are written to address teaching strategies for some of the key problem areas that emerged from our research.

As the owner of this book, you can use the Pretest to determine exactly which skills you need to target to pass the test. Once you have completed your study, you can determine whether you are ready to take the GED Science Test by taking an Official GED Practice Test—which follows Lesson 20. The GED Testing Service has developed this practice test as a predictor of the score that you will likely earn on the actual GED Science Test.

Remember that there are four other books in the **Keys to GED Success** series. These other books cover the remaining four GED Tests: Language Arts, Reading, Social Studies, Language Arts, Writing, and Mathematics. All titles in this series are available exclusively from Steck-Vaughn.

We wish you the best of luck on the GED Tests.

Sylvia E. Robinson

Executive Director
GED® Testing Service

September 2008

One Dupont Circle NW, Washington, DC 20036-1193
Telephone: 202/939.9490 Fax: 202/659.8875
www.acenet.edu www.GEDtest.org

STECK-VAUGHN

Keys to GED® SUCCESS

Science

www.SteckVaughn.com/AdultEd
800-531-5015

ISBN-10: 1-4190-5350-7
ISBN-13: 978-1-4190-5350-4

Printed in the United States of America.

1 2 3 4 5 6 7 8 9 022 15 14 13 12 11 10 09 08

Contents

KEY **This symbol indicates *GED® Key Skills* as identified by the GED Testing Service®.**

Using This Book

Keys to GED® Success: Science has been prepared by Steck-Vaughn in cooperation with the GED Testing Service®. This book focuses on the thinking and graphic interpretation skills needed to pass the GED Science Test.

This book also identifies the ***GED® Key Skills***, which are skills that the GED Testing Service® has pinpointed as those most often missed by test takers who come close to passing the GED Tests. For more information about these skills see *A Message from the GED Testing Service®* at the front of this book.

In this book, the ***GED Key Skills*** are identified by this symbol: **KEY**

It is recommended that students who are preparing to take the GED Tests follow this plan:

1. Take the Science Pretest.

While it is best to work through all the lessons in this book, students can choose to focus on specific skills. The *Science Pretest* assesses the 20 skills in this book. *The Pretest Performance Analysis Chart* on page 9 will help students to target the skills that need the most attention.

2. Work through the 4-page skill lessons in the book.

- The first page of each lesson provides an approach to the skill and to thinking through the questions. Students should carefully read the step-by-step thinking strategies and pay attention to the explanations of why the correct answers are right and why the wrong answer choices are incorrect.
- The second page of each lesson contains sample GED questions. Students should use the hints and the answers and explanations sections to improve their understanding of how to answer questions about each skill.
- The third and fourth pages of each lesson present GED practice questions that allow students to apply the skill to the same types of questions that they will see on the test.

Students should use the *Answers and Explanations* at the back of this book to check their answers and to learn more about how to make the correct answer choices.

3. Take the *Official GED® Practice Test Form PA: Science* in this book and analyze the results.

The half-length practice test at the end of this book is the Official GED Practice Test Form PA: Science–developed by the GED Testing Service®. Taking this test allows students to evaluate how well they will do on the actual GED Science Test.

Based on the results, test administrators can determine if the student is ready to take the actual test. Those students who are not ready will need more study and should use the other GED Science preparation materials available from Steck-Vaughn, which are listed at the back of this book and can be found at www.SteckVaughn.com/AdultEd.

4. Prior to taking the GED Science Test, take an additional Official GED Practice Test.

The more experience that students have taking practice tests, the better they will do on the actual test. For additional test practice, they can take the Full-Length Practice Test Form or any of the other Official GED Practice Tests available from Steck-Vaughn at www.SteckVaughn.com/AdultEd.

By using this book and the others in this series, students will have the information and strategies developed by both the GED Testing Service® and experienced adult educators, so that they can reach their goal—passing the GED Tests.

Teaching Tips

Below are suggested interactive teaching strategies that support and develop specific ***GED® Key Skills***.

Apply Scientific Principles (Science KEY Skill 5)

Discuss that on the GED Science Test, students may be given a scientific definition or principle, and they will need to recognize an example of that principle.

- Discuss a simple scientific definition. (For example: *Mixtures* are combinations of two or more different materials that are mixed physically but not chemically. The substances in a mixture retain their own properties and can be separated out by physical means, such as filtration, evaporation, distillation, etc.)
- Students come up with examples of the principle (For example: *muddy water* consists of dirt and water, *vinaigrette dressing* is oil and vinegar).
- Have a list of five items prepared. One of the items should fit the definition and the other four should not. Have students choose the correct example and support their answer. Then have them explain why the other four do not fit the definition. (For example, *fruit cake* is a mixture because you can separate the pieces of fruit out of the cake but *carbon dioxide, table salt, steel*, and *coffee* are not because the components can only be separated by chemical, not physical, processes.)

Combine Information from Text and Graphics (Science KEY Skill 18)

From a newspaper or news magazine, bring in science-related stories that are accompanied by one or more graphics. Use high-interest topics, such as health, weather, or environmental issues.

- Students read the story and brainstorm five main points from the story. Discuss: *What is the main point of this story? How does the writer use facts, details, or examples to make this point?*
- Students review the graphics. Discuss questions such as: *Why is this (graph, diagram, map, etc.) a part of this story? What does it relate to in this story? What point does it support? What did you learn from the graphic that expanded your understanding of the story?*

Use the Scientific Method (Science KEY Skill 20)

Discuss that the scientific method consists of the following stages: observing, forming a hypothesis, making a prediction, testing the prediction or gathering information, and drawing a conclusion. Suggest that this is something that scientists do, but it is also something that people do in daily life.

- Students brainstorm situations that they have observed in daily life. It could be related to something having to do with their home, their car, etc. Have them begin the following:
  - State something that has been observed.
  - Make a hypothesis about the reason for or cause of it.
  - Make a prediction.
- Outside of class they should:
  - Test the prediction or gather more information.
  - Draw a conclusion.

Discuss whether students' predictions were supported by the tests or the information that was gathered.

Common Misconceptions on the Science Test

GED Testing Service® research showed that test-takers often have difficulty with questions that are related to the issues listed below. Use your GED Science text or other materials to address these types of misconceptions.

- Preconceived notions, non-scientific beliefs, and conceptual misunderstanding
- Vocabulary: Common terms do not always have precisely the same meaning when used in a scientific context. Commonly used words and phrases, such as *energy* and *force, speed* and *acceleration, air* and *oxygen,* are often incorrectly used interchangeably.

Science Pretest

Directions

This pretest consists of 20 questions designed to measure how well you know the skills needed to pass the GED Science Test. There is one question for each of the 20 lessons in this book.

- Take the pretest and record your answers on the *Pretest Answer Sheet* found on page 117. Choose the one best answer to each question.
- Check your answers in the *Pretest Answers and Explanations* section, which starts on page 103. Reading the explanations for the answers will help you understand why the correct answers are right and the incorrect answer choices are wrong.
- Fill in the *Pretest Performance Analysis Chart* on page 9 to determine which skills are the most important for you to focus on as you work in this book.

---

1. A law of physics states that a body in motion will continue in motion until acted upon by a force of resistance. The greater the resistance, the more the motion is reduced, and eventually stopped. In which of the following circumstances is resistance to motion the greatest?

   (1) bowling ball rolling down an alley
   (2) tennis ball rolling on a grassy lawn
   (3) leaves blowing in the wind
   (4) rock falling off a cliff
   (5) car rolling on a flat road

2. As you watch distant fireworks, you notice that you hear the sound of the explosions a few seconds after you see the flashes of light. What can you infer from this observation?

   (1) Sound energy and light energy travel at the same speeds.
   (2) The fireworks are designed so that the sound occurs after the flash.
   (3) Light energy travels faster than sound energy.
   (4) Light energy travels in a straight line but sound energy does not.
   (5) Ears process information faster than eyes.

Question 3 refers to this graph.

Number of Legs of Selected Animal Groups

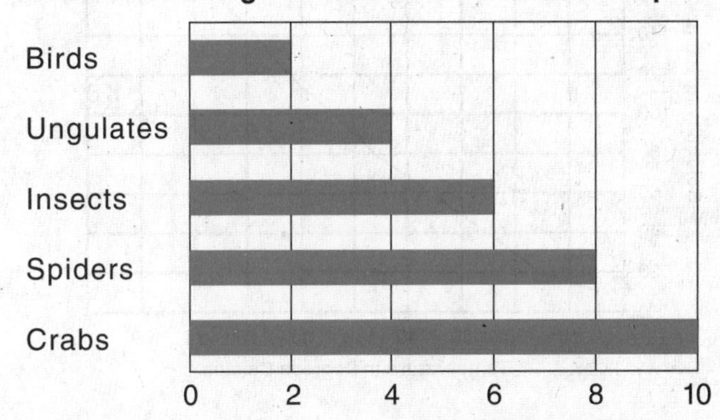

3. Based on the graph, which of the following statements is true?

   (1) All animals have even numbers of legs.
   (2) Spiders can run faster than insects because they have more legs.
   (3) Crabs have more complex bodies than insects.
   (4) Ungulates (hoofed mammals) have four legs.
   (5) Crabs and spiders have eight legs.

Questions 4 and 5 refer to this passage and graph.

The solubility of a compound in water varies with the temperature of the solution. The graph below shows the results of a student's lab experiment in which the solubility of two salts in water was measured at different temperatures.

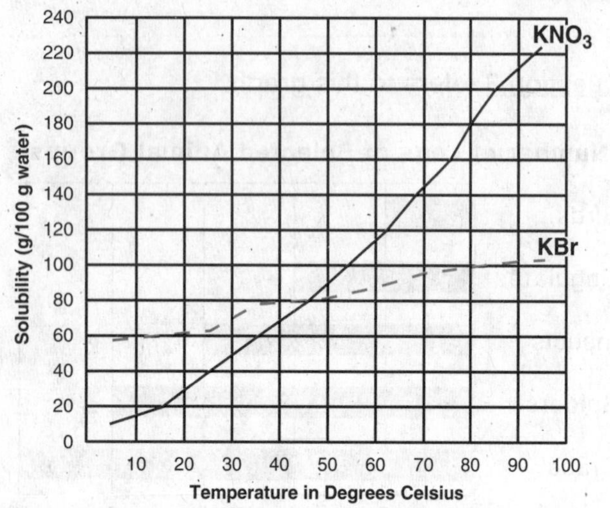

4. After collecting the data, the student formed a hypothesis that solubility depends on the negative ion of the salt. He predicted that the slope of the line representing solubility is steeper for salts containing metal ions and nitrate ions than it is for salts with the same metal ions and bromide ions. What should the student do next?

(1) Draw a conclusion based on the graphed data.
(2) Test the prediction using other nitrate and bromide salts.
(3) Report the result that nitrate salts have a steeper slope.
(4) Repeat the experiment with KNO_3 and KBr.
(5) Repeat the experiment using different temperatures.

5. Based on the data collected in the experiment, at what temperature is the solubility of the two salts identical?

(1) 6°C
(2) 38°C
(3) 46°C
(4) 50°C
(5) 88°C

Question 6 refers to following passage.

The bird called a robin in North America is completely different than the bird called a robin in China. Before Carolus Linnaeus devised his universal, Latin-based system for naming organisms in 1735, it was difficult for scientists to communicate accurately about organisms they were studying. Linnaeus's system classified organisms into orderly groups, with part of each organism's scientific name indicating the group to which it belonged. With this system, scientists can distinguish the North American robin from the Chinese robin because each has its own particular scientific name within the group *birds*. Linnaeus's classification system was a major contribution to the science of biology, and it is still used today.

6. Which of the following statements is implied by the above passage?

(1) Scientists could not study organisms prior to 1735.
(2) An accepted, orderly system aids scientific learning and communication.
(3) All scientific communication is in Latin.
(4) The scientific name for humans is *Homo sapiens*.
(5) Some organisms in the same group have the same scientific name.

Questions 7 and 8 refer to this table and passage.

In our solar system, eight planets orbit the sun. The average distance of a planet's orbit is measured in astronomical units (AU). One AU is the average distance between Earth and the sun, which is about 93 million miles, or 150 million kilometers. The table below shows the distance of each planet from the sun; its period of rotation on its axis, measured in hours; and its period of revolution around the sun, measured in Earth days.

Planet Data

Planet	Distance from sun (AU)	Rotation (hours)	Revolution (days)
Mercury	0.39	1048	88
Venus	0.72	5832	225
Earth	1.00	23.9	365
Mars	1.52	24.6	684
Jupiter	5.20	9.9	4,330
Saturn	9.54	10.6	10,800
Uranus	19.18	17.2	30,700
Neptune	30.06	16.1	60,200

7. Based on the information in the table, what is the average distance, in astronomical units, between Mars and the sun?

(1) 0.39
(2) 1.00
(3) 1.52
(4) 5.20
(5) 24.6

8. Which of the following statements is true based on the data in the table?

(1) The period of a planet's rotation increases as its distance from the sun increases.
(2) The period of a planet's revolution increases as its distance from the sun increases.
(3) The period of rotation is shorter for very large planets than for smaller planets.
(4) The period of revolution of a planet is unrelated to its distance from the sun.
(5) Rotation period decreases as revolution period increases.

Question 9 refers to this passage.

All living things respond to their environment in some way. If you clap your hands, your dog might immediately respond by looking to see from where the sound came. Plants respond to differences in light levels by growing toward a sunny windowsill instead of growing toward a dark room. Some bacteria move toward a food source by using tails that spin like propellers. During cold weather, you dress warmly before you go outside. All of an organism's responses to its environment together make up its behavior.

9. What is the main idea of the passage?

(1) Plants tend to grow toward light.
(2) Some living things respond to their environment and others do not.
(3) Bacteria and plants are examples of living things.
(4) People respond to cold weather.
(5) Living things exhibit behaviors as a response to their environment.

Question 10 refers to these diagrams.

Animal Cell

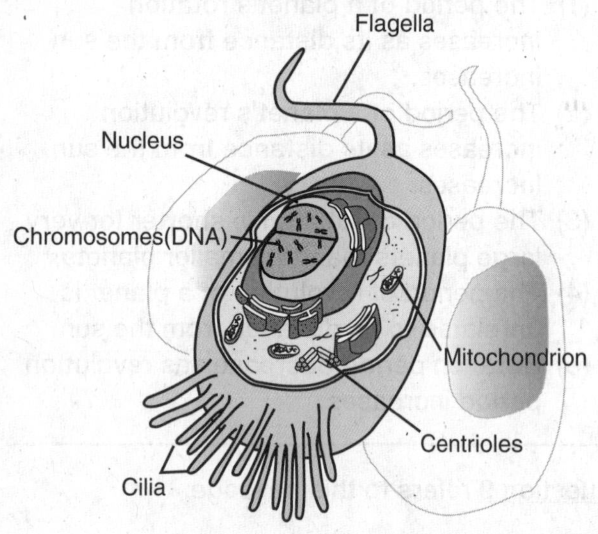

Virus Cell

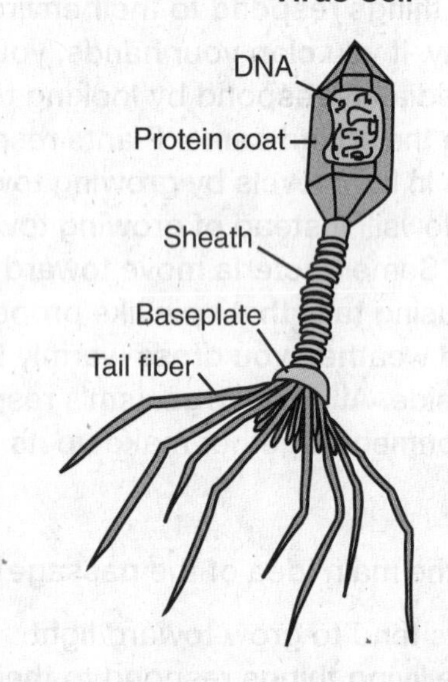

10. Based on the information in the diagrams, which of the following statements describes a difference between animal DNA and the DNA of a virus?

(1) The virus contains much more DNA than the animal cell.
(2) The DNA of the animal cell is more complex than the DNA of the virus.
(3) Animal DNA and virus DNA are composed of different types of bases.
(4) The DNA of animal cells is contained in the nucleus, but virus DNA is not.
(5) Virus cells have mitochondria, but animal cells do not.

Question 11 refers to following passage.

Some people argue that the statement that global warming can be caused by burning fossil fuels is clearly not supported by facts. When you look at the chemical reactions involved in the combustion of fuels, you find that the products are carbon dioxide and water, both of which are normal components of the atmosphere. Every time you take a breath, your body also sends carbon dioxide and water vapor into the atmosphere. If carbon dioxide, a chemical emitted by people breathing, were harmful to the environment, the planet would have become uninhabitable long ago. That fact demonstrates that burning fossil fuels cannot be harmful.

11. What is the faulty logic in this argument against fossil fuels causing global warming?

(1) When people breathe, they do not release carbon dioxide into the air.
(2) The carbon dioxide from burning fossil fuels is not the same as the carbon dioxide from breathing.
(3) Burning fossil fuels releases much more carbon dioxide into the atmosphere than breathing does, so the effects of the activities are not the same.
(4) The carbon dioxide from breathing is natural, but combustion of fossil fuels is an artificial source of carbon dioxide.
(5) The carbon dioxide from combustion causes more global warming because combustion occurs at higher temperatures than breathing.

Question 12 refers to this circle graph.

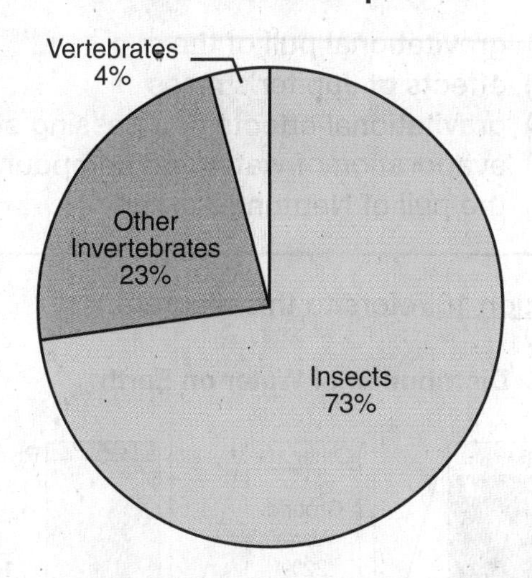

12. How does the number of vertebrate species compare with the total number of invertebrate species?

(1) about one-fourth fewer vertebrates than invertebrates
(2) about 3 times fewer vertebrates than invertebrates
(3) about 4 times as many vertebrates as invertebrates
(4) about 10 times as many vertebrates as invertebrates
(5) about 24 times fewer vertebrates than invertebrates

Question 13 refers to this passage.

Genetic modification is a relatively new technique that can be used in agriculture to give plants desirable traits. In one form of modification, the DNA in the cells of a plant is changed by inserting genes for the intended trait. Examples of this type of genetic modification include corn that resists the effects of certain herbicides, strawberries that can tolerate freezing weather, and wheat that manufactures compounds that kill parasites.

Although the technique has been used successfully to develop strains of plants with useful characteristics, there are potential problems. In some cases, the traits have been transferred to wild plants. This could lead to weeds that are able to resist insects or diseases that normally keep them in control. Some scientists think genetically modified plants should not be used in agriculture due to these risks. Others feel that the benefits are much greater than the risks.

13. Which of the following is an opinion stated in the passage?

(1) Genetic modification is a relatively new technique.
(2) Genetic modification changes specific traits in a plant.
(3) Genetic modification has not yet been successful in developing new plant strains.
(4) The risks of genetic modification are great enough that it should not be used.
(5) Modified traits can be transferred to other plants.

<u>Questions 14 and 15</u> refer to this passage.

A comet that passes near the sun is a spectacular sight. A comet is made of water and other compounds that have small molecules. These compounds evaporate, forming a large tail that reflects sunlight.

Comets are normally found far from the sun. Many famous comets, such as Halley's Comet, whose orbit brings it close to the sun every 76 years, come from the Kuiper Belt. This belt is a region beyond Neptune's orbit that is populated by many icy objects, including comets. When one of these objects drifts toward the sun, the gravitational pull of Jupiter or one of the other large planets can change its orbit. The comet then follows a path that brings it close to the sun periodically.

Other comets come from the Oort Cloud. This is the most distant part of the solar system, extending halfway to the nearest star. Comets in the Oort Cloud are too far away for their orbits to be changed by Jupiter. They can be affected by the gravitational pull of other stars, though. Occasionally, a passing star pushes one of these comets out of place. It then begins a million-year journey toward the inner part of the solar system, accelerated by the sun's gravity as it approaches. After passing the sun, these comets may begin a very long orbit or may leave the solar system altogether.

14. Which of the following is a summary of ideas presented in the passage?
    (1) Comets are part of the solar system.
    (2) Jupiter and the other large planets cause comets from the Kuiper Belt to move toward the sun.
    (3) Comets normally have very distant orbits, but some are pulled out of orbit by gravity and pass close to the sun.
    (4) A comet is made of water and other compounds with small molecules.
    (5) Most comets are destroyed when they crash into the sun and evaporate.

15. According to the passage, what causes comets from the Oort Cloud to begin moving toward the sun?
    (1) gravitational pull of the sun
    (2) effects of Jupiter's mass
    (3) gravitational effects of a passing star
    (4) evaporation of water and compounds
    (5) the pull of Neptune's gravity

<u>Question 16</u> refers to this diagram.

Distribution of Water on Earth

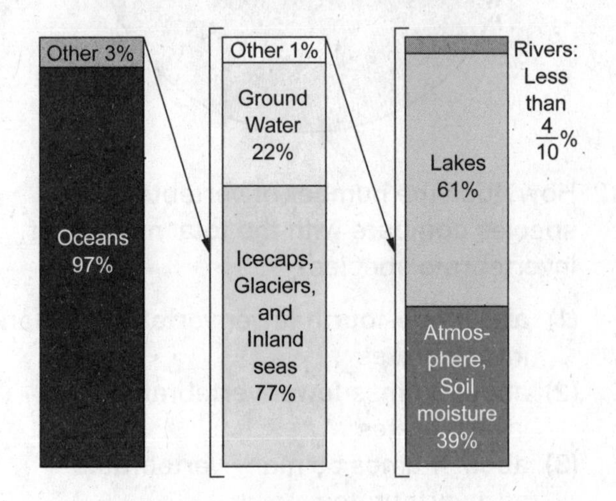

16. Which sentence below <u>best</u> restates the information in the diagram?
    (1) Water is distributed fairly evenly across the surface of Earth.
    (2) About 61% of water on Earth is in lakes and rivers.
    (3) The vast majority of Earth's water is in the oceans, and most of the rest is held in icecaps, glaciers, and inland seas.
    (4) Ground water is the main source of drinking water for people.
    (5) There is very little water in rivers and lakes because it evaporates rapidly.

Questions 17 and 18 refer to this passage and diagram.

Physicist Ernest Rutherford performed a number of experiments on the radiation emitted by radioactive materials. In one experiment, illustrated below, he passed a beam of radiation between two metal plates that had electrical charges. Three types of radiation were detected by the apparatus. Alpha rays were deflected toward the negative plate and beta rays toward the positive plate. The direction of the gamma rays was not affected by the charged plates. Rutherford concluded that alpha and beta rays were actually composed of electrically charged particles, while gamma rays had no electrical charge.

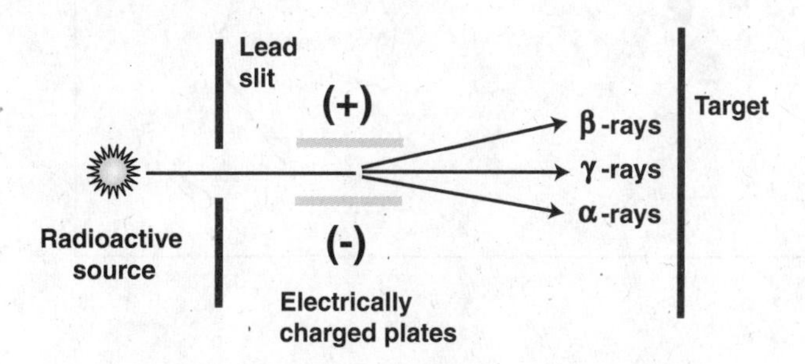

17. Based on the information in the passage and the diagram, what could Rutherford conclude about alpha and beta radiation?

(1) They both have positive charges, but their masses are different.
(2) Beta radiation and alpha radiation are negatively charged.
(3) Alpha radiation is positively charged, and beta radiation is negatively charged.
(4) The two types of radiation have different charges, but it cannot be determined which is positive and which is negative.
(5) The particles gain charges as they pass between the two plates.

18. What does the writer assume that the reader already knows in order to interpret the information and answer question 17?

(1) Lead stops all types of radiation.
(2) Opposite charges attract and like charges repel.
(3) Gamma rays are electromagnetic radiation, not particles.
(4) Alpha and beta rays were deflected exactly the same distance but in opposite directions.
(5) There are three different types of radiation.

Question 19 refers to this passage.

Before Galileo performed his experiments with gravity, it was believed that a heavier object fell faster than a lighter object. Although it is not certain that Galileo actually dropped objects from the Tower of Pisa, he did determine that objects fall to Earth at the same rate, regardless of mass. He also measured the rate of acceleration of falling objects and determined that gravity exerts a constant force on a falling object. As a result, the velocity increases constantly during the fall. If no other forces act on the object, the acceleration due to gravity is 9.8 m/s^2. That means that the velocity increases by 9.8 meters each second during the fall.

19. Which of the following statements is accurate and adequately supported by the information in the passage?

(1) Gravity is the only force acting on a falling object.
(2) The acceleration due to gravity is the same for any falling object.
(3) Galileo studied gravity by dropping objects from the Tower of Pisa.
(4) A helium balloon does not fall, so gravity does not apply a force on the balloon.
(5) A feather falls slower than a marble due to wind resistance.

Question 20 refers to following map.

Malaria is a mosquito-borne disease caused by protozoan parasites. The parasite's primary hosts are female mosquitoes. They inject humans with malaria when they feed.

People with malaria often experience fever, chills, tingling skin, and vomiting. Left untreated, they may develop severe complications and could die. According to the Center for Disease Control, each year 350-500 million cases of malaria occur worldwide.

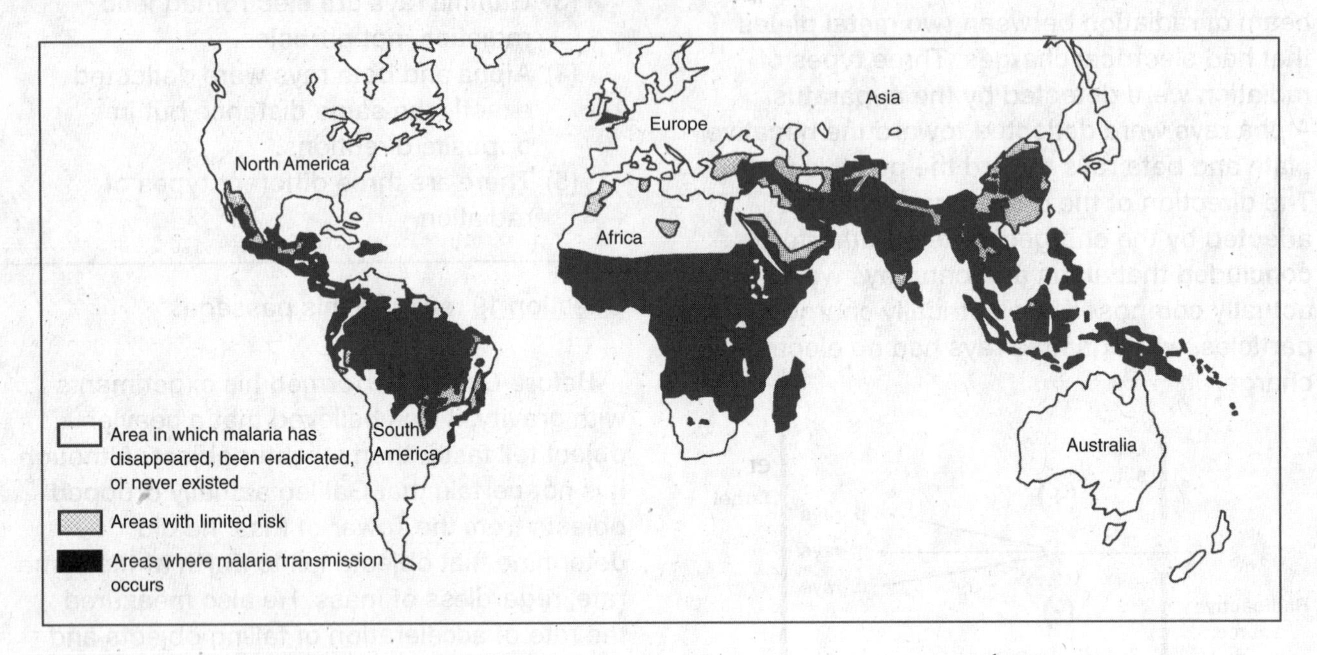

20. Which of the sentences below best restates information shown on the map?
    (1) Malaria is not a serious threat to most of the world's population.
    (2) The incidence of malaria is high in the United States and Canada because it has not been eradicated there.
    (3) Australia is likely to become an area of limited risk for malaria in the near future.
    (4) Tropical regions have the greatest occurrence of malaria.
    (5) Malaria transmission occurs most frequently in areas with large mosquito populations.

Joseph

Pretest Performance Analysis Chart

The following chart can help you to determine your strengths and weaknesses on the skill areas needed to pass the GED Science Test.

- Use the Science *Pretest Answers and Explanations* on pages 103–104 to check your answers.
- On the chart below
  - Circle the question numbers that you answered correctly.
  - Put a check mark (✓) next to the skills for which you answered the questions incorrectly.
  - Use the page numbers to find the lessons that you need to target as you work.

Question Number	Skills to Target (✓)	GED Science Skill Lessons	Page Numbers
9		**Skill 1:** Identify the Main Idea	10–13
20		**Skill 2:** Restate Information	14–17
14		**Skill 3:** Summarize Ideas	18–21
6		**Skill 4:** Identify Implications	22–25
1	1/30	**Skill 5:** Apply Scientific Principles	26–29
2	1/30	**Skill 6:** Make Inferences	30–33
13		**Skill 7:** Identify Facts and Opinions	34–37
18		**Skill 8:** Recognize Assumptions	38–41
15		**Skill 9:** Identify Causes and Effects	42–45
19		**Skill 10:** Assess Adequacy and Accuracy of Facts	46–49
8		**Skill 11:** Evaluate Information	50–53
11		**Skill 12:** Recognize Faulty Logic	54–57
7		**Skill 13:** Tables and Charts	58–61
3	1/30	**Skill 14:** Bar Graphs	62–65
5	1/30	**Skill 15:** Line Graphs	66–69
12		**Skill 16:** Circle Graphs	70–73
16		**Skill 17:** Diagrams	74–77
17		**Skill 18:** Combine Text and Graphics	78–81
10		**Skill 19:** Combine Information from Graphics	82–85
4	1/30	**Skill 20:** Use the Scientific Method	86–89

Skill 1

Comprehension

Identify the Main Idea

On the GED Science Test, you may have to answer questions in which you are asked to **identify the main idea.**

- The **main idea** is the central point of a passage or illustration.
- Other parts of the passage or illustration provide **supporting details.**

The passage below relates information about how materials are classified.

Read the passage. Choose the one best answer to the question.

Matter can be classified into three major classes based on its properties. The three classes are elements, compounds, and mixtures. Elements cannot be broken down into simpler substances by chemical means. Examples of elements are oxygen, iron, and silver. Compounds are substances in which elements are combined by chemical bonds. Carbon dioxide is an example of a compound. It is composed of carbon and oxygen. Mixtures are materials in which two or more substances are combined by physical means, rather than chemical bonds. A mixture may include elements, compounds, or a combination of elements and compounds. Air is a mixture of many gases, including oxygen, nitrogen, and carbon dioxide.

QUESTION: Which of the following statements best expresses the main idea of the passage?

(1) All matter occupies space and can be converted to energy.
(2) All matter can be classified as an element, a compound, or a mixture.
(3) Elements cannot be broken down into simpler substances by chemical means.
(4) Compounds are substances in which elements are combined by chemical bonds.
(5) Air is a mixture of many gases, including oxygen, nitrogen, and carbon dioxide.

EXPLANATIONS

STEP 1 To answer this question, ask yourself:

- What is this passage about? the three major classes of matter
- What is the question asking me to identify? the main idea of the passage

STEP 2 Evaluate all of the answer choices and choose the best answer.

(1) No. This is a definition of matter, but it is not the main idea of this passage.
(2) **Yes. The first two sentences of the passage state the main idea. The rest of the passage gives details that explain the three classes of matter.**
(3) No. This is a detail about elements, one of the three classes of matter, but it is not the main idea.
(4) No. This is a detail about compounds, one of the three classes of matter, but it is not the main idea.
(5) No. This is an example of a mixture, which is one of the three classes of matter.

ANSWER: (2) All matter can be classified as an element, a compound, or a mixture.

Practice the Skill

Try these examples. Choose the one best answer to each question. Then check your answers and read the explanations.

Scientists group similar organisms together in a system of classifications. The seven tiers of classification are shown in the diagram below, which progresses from the broadest category to the most specific. The classification of the grey wolf is given as an example.

Classifying Living Things

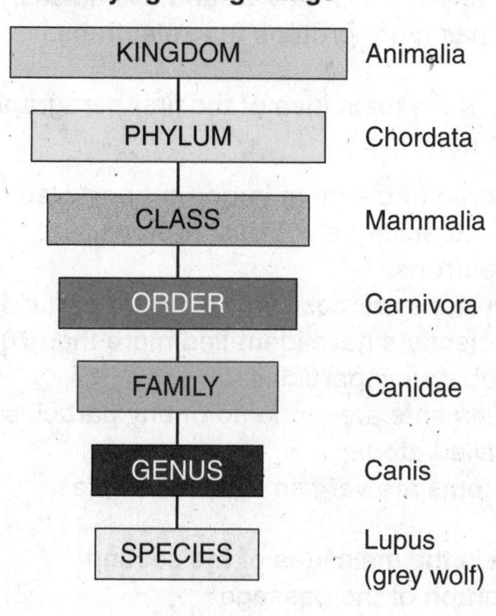

1. Which of the following statements best expresses the main idea of the diagram?

   (1) Grey wolves are called *Canis lupus*.
   (2) The species is the most specific level in the classification system.
   (3) Only animals are grouped in the classification system.
   (4) All living things are classified by scientists.
   (5) Living things are classified into seven different categories.

HINT **What is the purpose of the diagram?**

2. Which statement best expresses the main idea of the passage?

   (1) Scientists use a classification system to organize similar organisms.
   (2) Wolves are classified as *Mammalia*.
   (3) Classification is very important.
   (4) Grey wolves are the most important species to classify.
   (5) The seven tiers progress from the broadest to the most specific.

HINT **What is the main point of the passage that introduces the diagram?**

Answers and Explanations

1. (5) Living things are classified into seven different categories.
Option (5) is correct because the title of the diagram is "Classifying Living Things," and the diagram shows seven categories.

The main idea of the diagram is the classification system, not the classification of wolves (option 1). Option (2) is only a detail of the graph. The title indicates that the system applies to all living things, not just animals (option 3). Although the passage states that scientists use the classification system (option 4), this information is related only to the passage and not related to the diagram.

2. (1) Scientists use a classification system to organize similar organisms.
Option (1) is correct because the main point of the passage is to introduce the classification system used by scientists to organize similiar organisms.

Option (2) is a detail of the diagram, but not of the passage. The importance of classification (option 3) is not discussed in the passage. Grey wolves (option 4) are mentioned as the example to be shown in the diagram, but there is no indication that they are the most important idea, and this is not the main idea of the passage. Option (5) is not the main idea of the passage; it is a supporting detail.

GED Practice

Identify the Main Idea

Directions: Choose the one best answer to each question.

Questions 1 through 3 refer to the following passage and diagram.

All elements are made up of tiny particles called atoms. An atom is the smallest particle of an element. Atoms are very small and complex, containing even smaller parts called subatomic particles. More than 70 subatomic particles have been identified by scientists.

Three of the most important particles in atoms are electrons, protons, and neutrons. Electrons are negatively charged particles. They move around the center, or nucleus, of the atom. Protons and neutrons located in the nucleus are grouped closely together. Protons are positively charged particles, and neutrons are particles without a charge. The arrangement of electrons, protons, and neutrons in a helium atom is shown in the diagram below.

Helium Atom

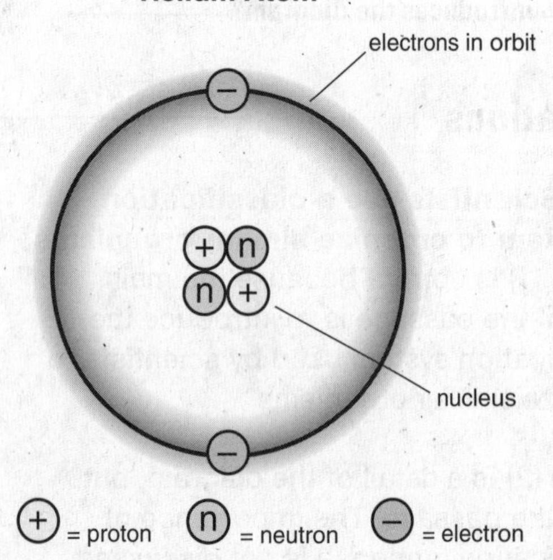

1. Based on the diagram, which statement is true of a helium atom?
   (1) It has four neutrons.
   (2) It has four particles in the nucleus.
   (3) One electron moves around its nucleus.
   (4) Two protons move around its nucleus.
   (5) It has more protons than electrons.

2. What is the main idea of the first paragraph of the passage?
   (1) Three of the most important particles in atoms are electrons, protons, and neutrons.
   (2) Protons are positively charged particles.
   (3) Scientists have identified more than 70 subatomic particles.
   (4) Elements are made up of tiny particles called atoms.
   (5) Atoms are very small and complex.

3. What is the main idea of the second paragraph of the passage?
   (1) Atoms are very small and contain tiny parts called subatomic particles.
   (2) Protons and neutrons are grouped closely together in the nucleus.
   (3) Three of the most important particles in atoms are electrons, protons, and neutrons.
   (4) A helium atom contains an arrangement of electrons, protons, and neutrons.
   (5) Electrons are negatively charged particles.

Questions 4 and 5 refer to the following passage and diagram.

Aspirin is very effective in treating fever and inflammation. However, the active ingredient in aspirin, salicylic acid, can irritate the stomach lining, and extensive use can lead to bleeding and stomach ulcers.

PolyAspirin is a new drug that offers the benefits of aspirin without its unwanted side effects. PolyAspirin is made up of about 100 aspirin molecules linked together like beads in a necklace. This structure allows the drug to pass through the stomach in one piece. It breaks down in the intestine, where it is absorbed and sent to its target.

PolyAspirin

■ = breaks down into salicylic acid

4. What is the main idea of the passage?
   (1) Aspirin can irritate the stomach.
   (2) PolyAspirin works like aspirin but is safer for the stomach.
   (3) PolyAspirin breaks down in the intestine.
   (4) Salicylic acid is the main ingredient of aspirin.
   (5) Aspirin is used to treat fevers and inflammation.

5. What is the main idea of the diagram?
   (1) PolyAspirin is a chain that breaks down into smaller molecules, including salicylic acid.
   (2) All of the molecules linked together in PolyAspirin are the same.
   (3) PolyAspirin breaks down in the intestine.
   (4) PolyAspirin will replace aspirin use in the future.
   (5) Some parts of a PolyAspirin molecule do not turn into salicylic acid.

Question 6 refers to the following passage.

One of the most powerful forces that shapes the surface of Earth is a moving glacier. A glacier forms when snow and ice accumulate over many years without melting. As more and more layers of ice build, the total weight of the ice causes the glacier to move outward. If the glacier forms on a mountain, the ice moves downhill.

Glaciers move very slowly. The fastest glacier movement ever recorded was just under 8 miles per year. Even so, they can cause significant changes in a landscape. For example, in Yosemite National Park there are cliffs that are greater than a mile high. These cliffs formed when glaciers flowed through the Yosemite Valley millions of years ago. They carried away rocks from the mountains on both sides of the valley, leaving a huge, deep scar in the land. Millions of years later, we can still see the power of those glaciers.

6. What is the main idea of the passage?
   (1) Glaciers are made of ice and snow.
   (2) Most glaciers move about 8 miles per year.
   (3) Yosemite National Park has cliffs more than a mile high.
   (4) Millions of years ago, glaciers were larger than they are today.
   (5) Glaciers can cause massive changes in the landscape.

TIP

To understand the main idea of a passage or illustration, ask yourself the following:
- **What is the main point?**
- **What is the idea that the details and examples support?**

Answers and explanations start on page 105.

Skill 2

Comprehension

Restate Information

On the GED Science Test, you may have to answer questions in which you are asked to **restate information.**

- To **restate** an idea, repeat it in a different way.
- The idea stays the same, but the facts are **rearranged** or stated in another way.

The passage below presents information about cells.

Read the passage. Choose the one best answer to the question.

Cells are the building blocks of all living things. A cell is the smallest living unit that can carry out the functions of life: growth, feeding, and reproduction. Some organisms, such as bacteria, consist of a single cell. These cells reproduce by dividing to form two new cells that have the same characteristics as the original cell.

More complex organisms, such as insects, plants, and even humans, have many cells that work together. These organisms have a huge number of cells that are able to perform specialized functions. For example, your muscle cells, bone cells, and red blood cells all have specific jobs that are necessary to keep you alive. Like single-celled organisms, the specialized cells can grow, feed, and reproduce, even though they are not separate organisms.

QUESTION: Which statement below best restates the description of a cell in the passage above?

(1) the smallest unit of living and nonliving matter
(2) a living organism that can survive on its own
(3) the smallest part of an organism that can carry out life functions
(4) the basic building block of bacteria
(5) a living unit that always specializes in a particular function

EXPLANATIONS

STEP 1 To answer this question, ask yourself:

- What is the key idea of the passage? cells are the building blocks of life
- What is the question asking me to do? restate the information about cells

STEP 2 Evaluate all of the answer choices and choose the best answer.

(1) No. Cells are the building blocks of life, so they cannot be nonliving matter. This choice is not correct.
(2) No. This description only describes single-cell bacteria, which is a very specific type of cell. The description is too specific to be correct.
(3) **Yes. This description includes all types of cells. It is an accurate restatement of the description of a cell from the second sentence.**
(4) No. This does not restate information about cells in general, but of a specific type of cell: bacteria.
(5) No. This description excludes single-cell organisms and is, therefore, too specific to be an accurate restatement.

ANSWER: (3) the smallest part of an organism that can carry out life functions

Practice the Skill

Try these examples. Choose the one best answer to each question. Then check your answers and read the explanations.

Questions 1 and 2 refer to this passage and graph.

The surface of the sun is very active. Changes in the sun's magnetic field create regions known as sunspots. Sunspots are cooler than the surrounding surface. During periods of high sunspot activity, solar flares also tend to occur. Solar flares fling a huge number of charged particles into space. When these particles reach Earth, they can interfere with satellites and communication systems.

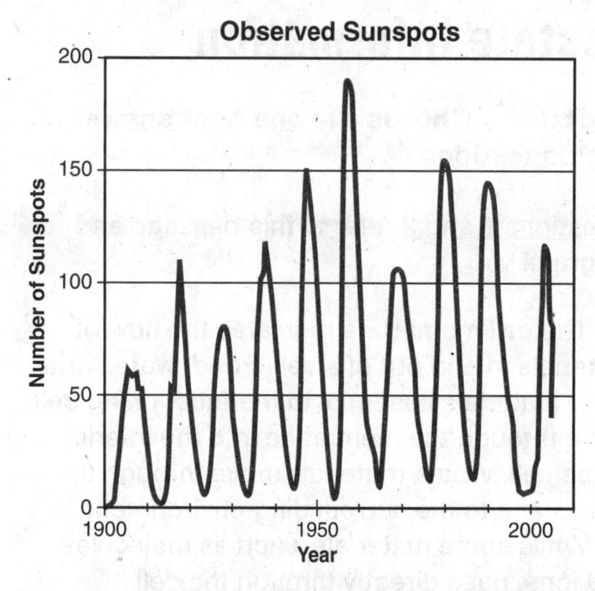

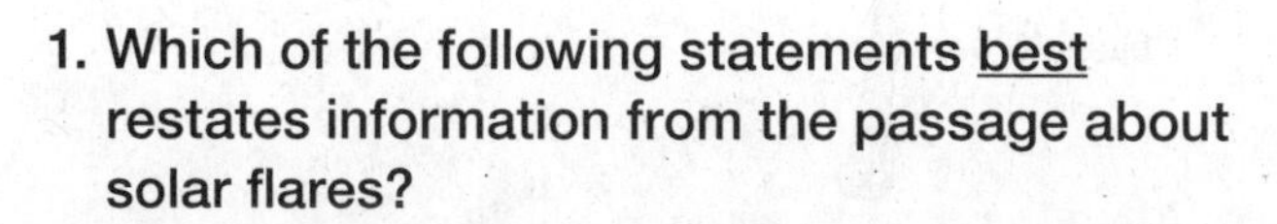

1. Which of the following statements best restates information from the passage about solar flares?
   (1) Solar flares cause magnetic activity on the sun.
   (2) Solar flares are a source of charged particles that reach Earth.
   (3) Solar flares usually occur in summer.
   (4) Solar flares are cool regions on the sun.
   (5) Solar flares extend from the sun to Earth.

HINT **What happens during solar flares?**

2. Which sentence best restates the graphical information about sunspots?
   (1) Solar flares occur a few times per year.
   (2) Sunspot activity cannot be predicted.
   (3) Sunspots are cool regions on the sun's surface.
   (4) The number of sunspots increases every year.
   (5) The number of sunspots varies in cycles about every ten years.

HINT **Use the information on each axis to interpret information on the graph.**

Answers and Explanations

1. (2) Solar flares are a source of charged particles that reach Earth.

Option (2) is correct because the passage states that charged particles from solar flares interfere with communications when they reach Earth. Option (2) is an accurate restatement.

Solar flares are the result, not the cause, of magnetic activity (option 1). The passage does not state when flares occur (option 3) or how far they extend (option 5). Cool regions on the sun are sunspots, not solar flares (option 4).

2. (5) The number of sunspots varies in cycles about every ten years.

Option (5) is a correct restatement because the graph clearly shows an approximate ten-year cycle of increase and then of decrease in sunspot activity.

The graph shows sunspot activity, not solar flares (option 1). It is possible to make predictions from the graph, so option (2) is not correct. The description of sunspots as cool regions (option 3) is from the passage, not the graph. The graph shows the activity increases and decreases; it does not only increase (option 4).

GED Practice

Restate Information

Directions: Choose the one best answer to each question.

Questions 1 and 2 refer to this passage and diagram.

The cell membrane regulates the flow of materials in and out of a cell. Food, water, and other materials essential to the functioning cell move through the membrane into the interior of the cell. Waste materials move through the membrane to the surrounding environment.

While some materials, such as molecules and ions, pass directly through the cell membrane itself, other materials are too large to do so. The diagram below shows two ways that large particles cross the cell membrane.

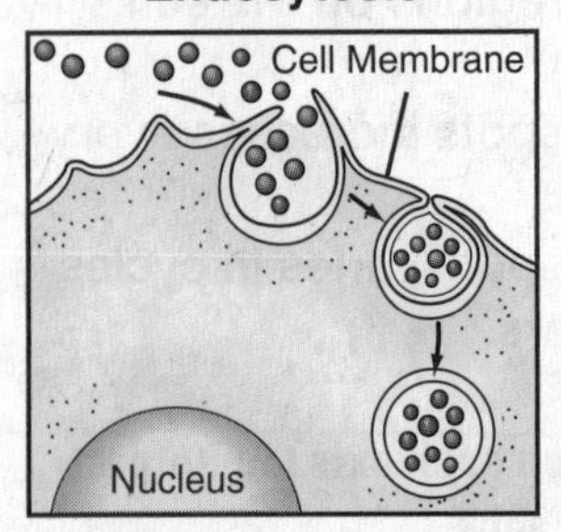

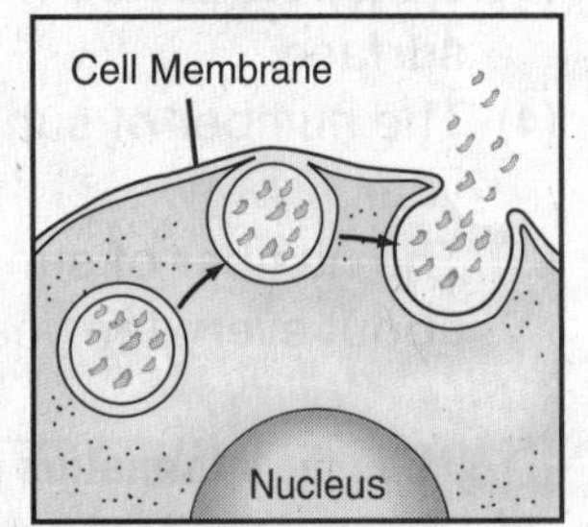

1. Which kind of material is likely to exit a cell via exocytosis?

   (1) food
   (2) water
   (3) nucleus
   (4) large waste particles
   (5) oxygen

2. Which of the following best restates the first step of endocytosis?

   (1) The cell membrane surrounds large particles.
   (2) The cell membrane absorbs small particles directly.
   (3) The cell expels small particles.
   (4) The cell expels large particles.
   (5) The cell membrane prevents particles from entering the cell.

Question 3 refers to this passage and diagram.

The acceleration (a) of an object is a function of its mass (m) and the force (F) that acts on the object as shown in the diagram below.

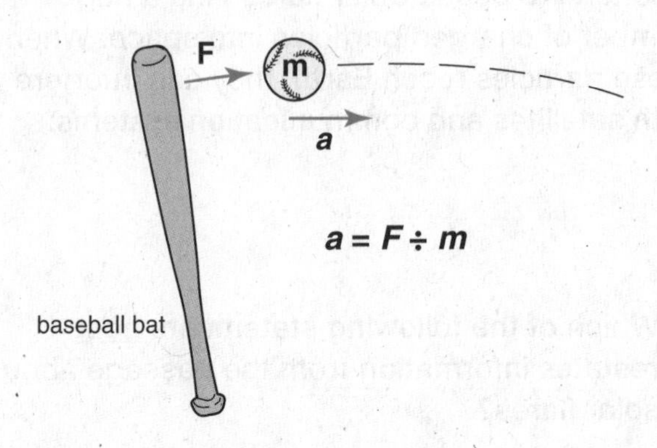

3. Which of the following correctly restates the relationship between acceleration, force, and mass?

   (1) A ball accelerates when it is hit by a bat.
   (2) A larger ball is easier to hit than a smaller ball.
   (3) Acceleration is the product of force and mass.
   (4) Acceleration can be calculated by dividing the force by the mass.
   (5) Acceleration occurs in the same direction as that of the force.

TIP

Some GED Science Test questions will ask you to look at a diagram and to describe what you see in words. Use the word labels on the diagram to locate and restate specific information.

Questions 4 and 5 refer to this passage and diagram.

Earth receives energy from the sun in the form of light. Some of this energy goes back into space as infrared radiation. Certain gases in our atmosphere, called greenhouse gases, trap energy just like the glass of a greenhouse. This is called the greenhouse effect. Without it, Earth would be too cold to support life.

Some human activities, such as burning fossil fuels, add greenhouse gases to the atmosphere. Scientists are concerned that the excess greenhouse gases will cause a rapid increase in Earth's temperature, a phenomenon called global warming. Global warming can cause changes in climate, leading to increased precipitation and rising sea levels.

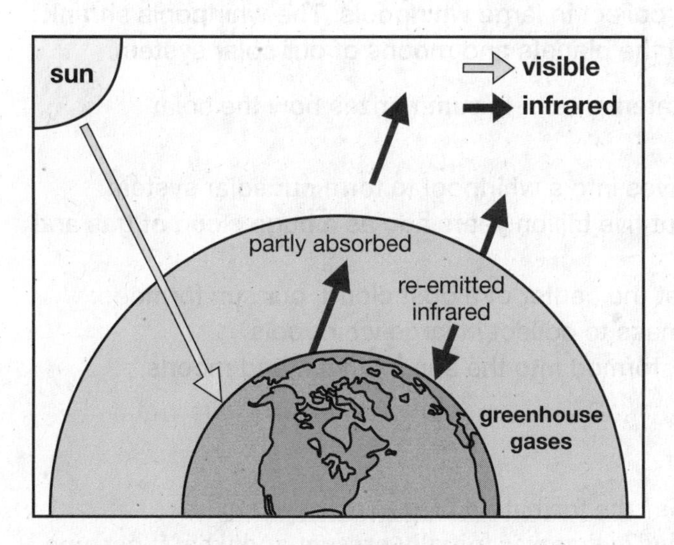

4. Which is the best restatement of the information in the first paragraph and the diagram?
   (1) Greenhouse gases are harmful to Earth.
   (2) Energy comes from the sun.
   (3) The greenhouse effect makes Earth warmer.
   (4) Global warming can cause climate changes.
   (5) Greenhouse gases trap energy from the sun within the earth's atmosphere.

5. Which of the following statements supports the concerns about global warming as described in the passage?
   (1) Global warming supports life.
   (2) Global warming can cause changes in the weather.
   (3) Without global warming, all the sun's energy is reflected back into space.
   (4) Global warming reflects too much light.
   (5) Burning of fossil fuels is reducing the amount of oxygen available for humans.

Question 6 refers to this passage and diagram.

Scientists organize all the known elements in order of increasing atomic number on a table known as the periodic table. On the periodic table, elements are arranged in horizontal rows called periods and in vertical columns called groups. Elements in the same period have electrons located in the same energy level, or electron shell. Elements in the same group have the same number of electrons in their outer shell, as well as similar chemical and physical properties.

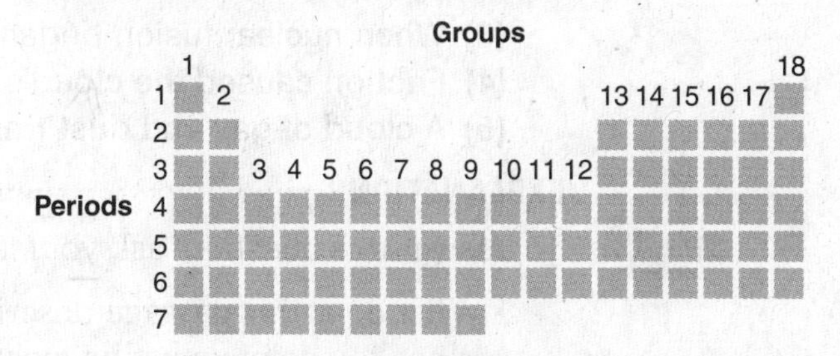

6. How could you best restate the information in the passage and diagram?
   (1) The periodic table contains seven periods and eighteen groups.
   (2) Each square of the periodic table represents an element, and the squares are organized based on structures and properties of the elements.
   (3) The periodic table includes all of the known information about each element.
   (4) Elements in a group of the periodic table have the same number of electrons in their outer shell.
   (5) All matter is made of about 100 different elements, which are shown on the periodic table.

Answers and explanations start on page 105.

Skill 3

Comprehension

Summarize Ideas

On the GED Science Test, you may have to answer questions in which you are asked to **summarize the ideas** presented in a passage or illustration.

- To **summarize,** restate the main idea and key points in simple words.
- A **summary** is shorter than the original information and has less detail, but it is factually correct.

The passage below relates information about how the solar system formed.

Read the passage. Choose the one best answer to the question.

The solar system began about five billion years ago as a huge cloud of gas and dust. Shock waves from an outside source caused the cloud to collapse inward, become dense, and spin. Gas particles collided and the center of the cloud grew hotter. Nuclear fusion began, and a star, our sun, formed at the center of the cloud. About ten percent of the cloud formed a plate-shaped disk surrounding the sun. Friction caused the cloud's mass to collect in large whirlpools. The whirlpools shrank into dense masses that later formed the planets and moons of our solar system.

QUESTION: Which of the following statements best summarizes how the solar system formed?

(1) A cloud of gas sent shockwaves into a whirlpool to form our solar system.
(2) The solar system began about five billion years ago as a huge cloud of gas and dust.
(3) When nuclear fusion began at the center of a dust cloud, our sun formed.
(4) Friction caused the cloud's mass to collect in large whirlpools.
(5) A cloud of gas and dust transformed into the sun, planets, and moons.

EXPLANATIONS

STEP 1 To answer this question, ask yourself:

- What does this passage describe? the formation of the solar system
- What key points must be included? what was initially present and what it became

STEP 2 Evaluate all of the answer choices and choose the best answer.

(1) No. This statement is not true and confuses facts from the passage.
(2) No. This is only a detail of how the solar system formed. It is too specific to be a summary.
(3) No. This describes the formation of the sun, but not the solar system.
(4) No. This is only a supporting detail of how the solar system began. It does not include enough information to be a summary.
(5) **Yes. This is a factually correct, brief statement of the formation of the solar system.**

ANSWER: (5) A cloud of gas and dust transformed into the sun, planets, and moons.

Practice the Skill

Try these examples. Choose the one best answer to each question. Then check your answers and read the explanations.

Questions 1 and 2 refer to this passage and diagram.

When the moon or Earth passes through the other's shadow, an eclipse occurs. Two types of eclipses involving the sun, the earth, and the moon are shown in the diagrams to the right. The plane of the moon's orbit is tilted at a 5° angle to that of Earth's orbit. As a result, eclipses do not occur each time the moon completes one revolution around the earth. Eclipses only occur when the moon crosses the plane of Earth's orbit.

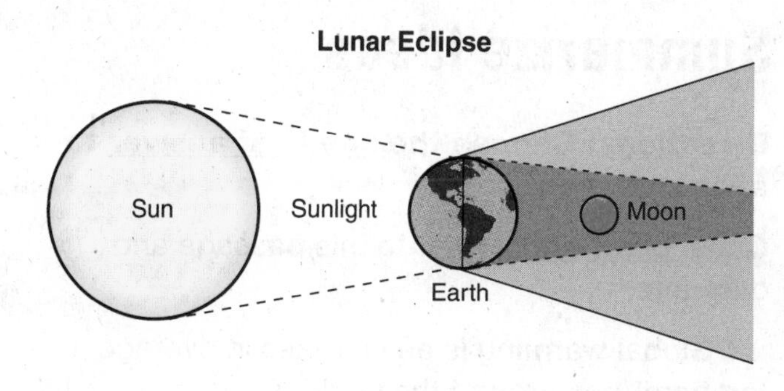

1. Which sentence best summarizes what causes a solar eclipse?
   (1) The moon casts a shadow on Earth.
   (2) The sun and the moon are on the same side of Earth.
   (3) Earth casts a shadow on the moon.
   (4) The moon's orbit is tilted at a 5° angle to Earth's orbit.
   (5) The moon crosses the plane of Earth's orbit and passes into its shadow.

HINT **What is the key idea illustrated in the diagram of a solar eclipse?**

2. According to the diagram, what happens when the moon passes through Earth's shadow?
   (1) The sun appears dark from Earth.
   (2) The moon blocks the sun's light.
   (3) A solar eclipse occurs.
   (4) A lunar eclipse occurs.
   (5) Earth completes one revolution around the sun.

HINT **What kind of eclipse occurs when the moon passes on the opposite side of the earth from the sun?**

Answers and Explanations

1. (1) The moon casts a shadow on Earth. Option (1) is correct because it correctly summarizes that the moon casts a shadow on Earth during a solar eclipse.

Although the sun and moon are on the same side of Earth (option 2) during an eclipse and the moon's orbit does tilt relative to Earth's orbit (option 4), these are only details, and not the cause of a solar eclipse. The other two choices (options 3 and 5) are incorrect statements based on the information in the passage and the diagram.

2. (4) A lunar eclipse occurs. Option (4) is correct because the moon passes through Earth's shadow during a lunar eclipse according to the diagram.

The sun appearing dark from Earth (option 1) and the moon blocking sunlight (option 2) are descriptions of a solar eclipse, not a lunar eclipse. A solar eclipse (option 3) occurs when the moon passes between the sun and Earth, not in the shadow of the sun. The diagram does not address the earth's revolution around the sun (option 5).

GED Practice

Summarize Ideas

Directions: Choose the one best answer to each question.

Questions 1 and 2 refer to this passage and diagram.

Global warming is an increase in average temperatures around the world. Carbon dioxide and other gases in the earth's atmosphere trap heat from the sun and keep the climate suitable for life. When humans burn coal, oil, and gasoline, more carbon dioxide is released into the atmosphere. The graph below shows the carbon dioxide concentration in the atmosphere and the change in global temperatures over time.

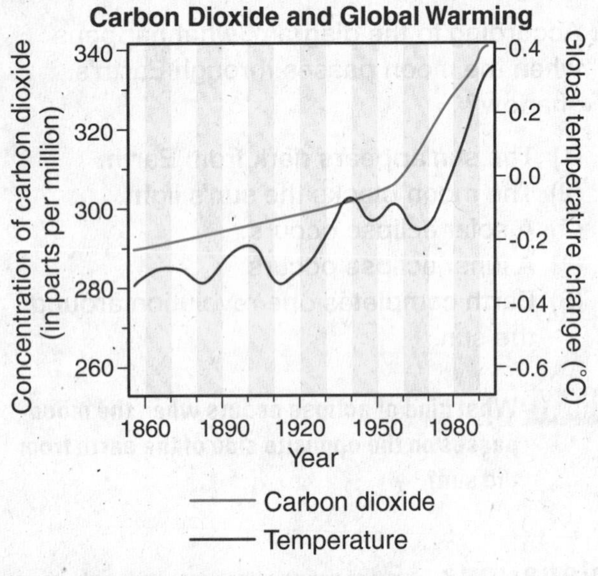

1. Which sentence is the best summary of the information in the graph?
   (1) Carbon dioxide levels increased steadily.
   (2) Carbon dioxide levels decreased.
   (3) Global temperatures fluctuated greatly.
   (4) Global temperatures decreased and carbon dioxide levels increased.
   (5) Carbon dioxide levels and global temperatures generally increased.

2. According to the data, what is most likely to happen if humans burn coal, oil, and gasoline at an increasing rate?
   (1) Carbon dioxide will decrease.
   (2) Gases in the earth's atmosphere will stop trapping the sun's heat.
   (3) Global temperatures will go down.
   (4) Global warming will end.
   (5) Carbon dioxide levels and global temperatures will continue to rise.

Question 3 refers to this table.

Inner Planets	
Planet	Average surface temperature (°C)
Mercury	350
Venus	460
Earth	20
Mars	−23

3. What is the best summary of the data in the table?
   (1) Venus has the hottest temperature.
   (2) Mars has the coldest temperature.
   (3) Earth is warmer than Mars but colder than Mercury.
   (4) The four planets have temperatures ranging from −23°C to 460°C.
   (5) Earth has an average surface temperature of 20°C.

TIP

Sometimes information will be presented in a diagram, map, graph, or chart. Read graphics carefully to determine the best summary of the information.

Questions 4 and 5 refer to this passage and illustration.

Deoxyribonucleic acid (DNA) is the molecule that makes up genes. DNA is found in the nucleus of the cell, and it contains the coded instructions for the organism's growth and development. It also contains the code for traits that are passed from one generation to the next.

DNA is made of two spiral threads, connected by crosspieces, in a shape called a double helix. Genetic information is stored in a code based on the order of four types of molecules. During cell division, DNA splits in half lengthwise, forming two threads. Each half of the DNA molecule copies the missing thread to form a new, complete double helix.

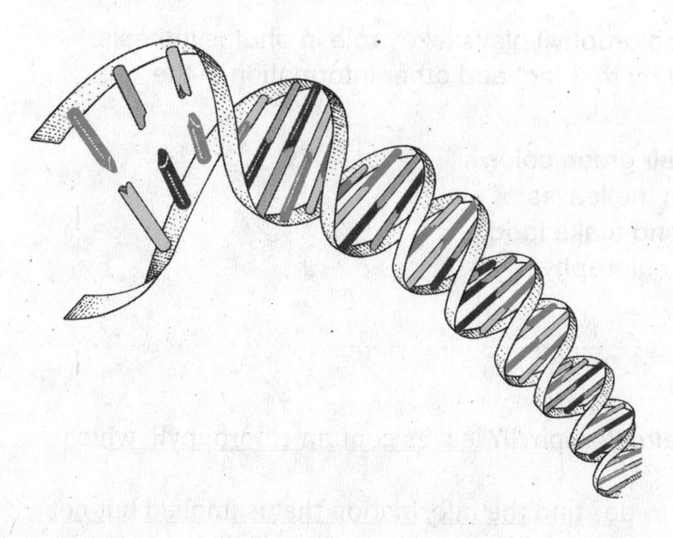

4. Which statement best summarizes the information in the passage?
   (1) The shape of a DNA molecule is a double helix.
   (2) DNA molecules store instructions for cell growth and genetic information in a pattern of molecules.
   (3) Animals cannot live without DNA in their cells.
   (4) A DNA molecule can divide into two parts by separating along the central linkages.
   (5) Cells divide when an organism reproduces.

5. Which summary best describes the information in the illustration?
   (1) DNA is important for cell reproduction.
   (2) All plant and animal cells have DNA.
   (3) A DNA molecule is very long.
   (4) A DNA molecule can divide into two parts by separating along the central linkages.
   (5) DNA molecules store genetic information.

Question 6 refers to this passage and graph.

The graph below shows the predicted effects of limiting the use of CFCs, chlorine-containing chemicals.

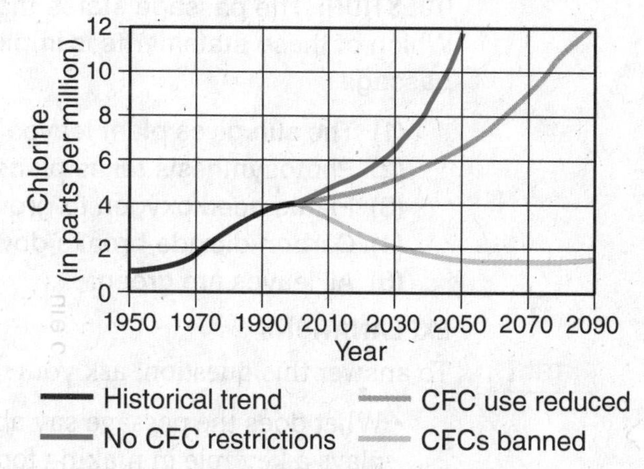

6. Which summary best describes the information in the graph?
   (1) Banning CFCs will preserve the ozone layer.
   (2) Banning CFCs will have no effect on the atmosphere.
   (3) Only banning CFCs will reduce the amount of chlorine in the atmosphere.
   (4) No restrictions on CFCs will decrease the amount of chlorine in the atmosphere.
   (5) The amount of chlorine in the atmosphere will increase regardless of the use of CFCs.

Answers and explanations start on page 105.

Skill 4

Comprehension

Identify Implications

On the GED Science Test, you may have to answer questions about implications.

- An **implication** is an idea or fact that is not directly stated in words. It must be inferred, or figured out logically.
- An implication gives enough information to imply, or suggest, an idea or fact **logically.**

The passage below includes implied information about how plants make food.

Read the passage. Choose the one best answer to the question.

The leaves of a plant contain a green substance called chlorophyll, which plays a key role in making the plant's food. In this process, called photosynthesis, the sun provides energy to break down water into hydrogen and oxygen, which combines with carbon dioxide to form simple sugars. The plant uses the sugars as food and to form other substances, such as starches.

QUESTION: The passage states that chlorophyll plays a key role in photosynthesis. Which of these statements is implied by that fact and other information in the passage?

(1) The sun gives plant leaves their green color.
(2) Photosynthesis takes place in the leaves of plants.
(3) Plants need oxygen to grow and make food.
(4) Carbon dioxide breaks down chlorophyll.
(5) All leaves are green.

EXPLANATIONS

STEP 1 To answer this question, ask yourself:

- What does the passage say about chlorophyll? leaves contain chlorophyll, which plays a key role in making food
- What is the question asking me to do? find the information that is implied but not stated

STEP 2 Evaluate all of the answer choices and choose the best answer.

(1) No. This passage gives no implication that the sun causes leaves to be green.
(2) **Yes. Chlorophyll is found in leaves and it plays a key role in photosynthesis, so the passage implies that photosynthesis takes place in leaves.**
(3) No. This statement contradicts the passage. It is not implied.
(4) No. Chlorophyll and carbon dioxide are involved in different parts of photosynthesis. The passage gives no indication that carbon dioxide breaks down chlorophyll.
(5) No. Although chlorophyll is green, the statement does not imply that all leaves are green.

ANSWER: (2) Photosynthesis takes place in the leaves of plants.

Practice the Skill

Try these examples. Choose the one best answer to each question. Then check your answers and read the explanations.

Questions 1 and 2 refer to the passage and diagram.

Our galaxy, the Milky Way, is a spiral galaxy. The center of the galaxy contains hundreds of millions of closely-packed stars in what is called the nuclear bulge. Stars outside the bulge lie in a thin disk about 100,000 light years across and a few hundred light years thick. Spiral arms that contain the galaxy's youngest, brightest stars wrap around the nuclear bulge. A sphere of stars, called the halo, spreads out in all directions around the disk.

Milky Way Galaxy

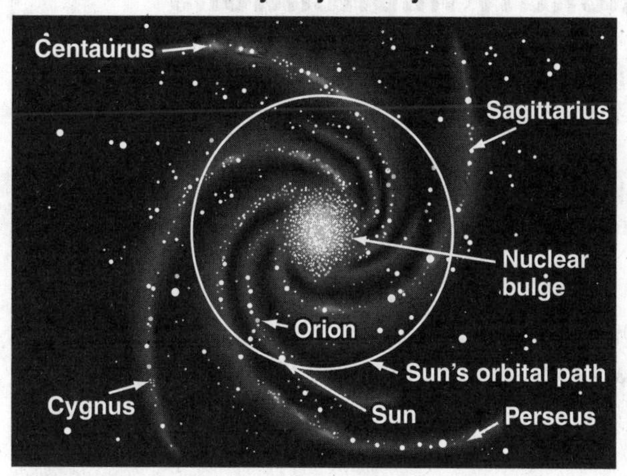

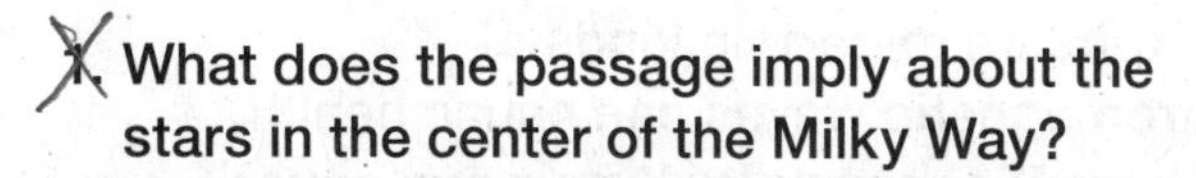

1. What does the passage imply about the stars in the center of the Milky Way?
   (1) There are hundreds of millions of them.
   (2) They are older than the stars in the spiral arms.
   (3) Earth is a part of the nuclear bulge.
   (4) The Milky Way is a spiral galaxy.
   (5) They are the brightest stars in our galaxy.

HINT How do the stars in the center compare to those in the spiral arms?

2. What can be implied about the stars in the spiral arms of the galaxy from the diagram?
   (1) They revolve around the nuclear bulge.
   (2) They are larger than the stars in the bulge.
   (3) They are the youngest stars.
   (4) They move between spiral arms.
   (5) Our sun is older than most stars in the bulge.

HINT What statement can you infer to be true by looking at the diagram?

Answers and Explanations

1. (2) They are older than the stars in the spiral arms.
Option (2) is correct. The passage states that stars in the spiral arms are youngest, implying that stars in the center are older.

There are hundreds of millions of stars (option 1), but this is a direct statement from the passage, not an implication. There is no information to imply that Earth is in the nuclear bulge (option 3). The Milky Way is a spiral galaxy (option 4), but this is the main idea, not an implication. The passage states that stars in the spiral arms are the brightest, which implies that the stars in the center are not the brightest, so option (5) is wrong.

2. (1) They revolve around the nuclear bulge.
Option (1) is correct because the illustration shows the orbit of the sun, which is a star. This implies that the stars in the arms revolve around the center.

That they are the youngest stars (option 3) is stated in the passage, not in the diagram. The size of the stars (option 2) and motion relative to the arms (option 4) are not indicated. Option (5) contradicts the passage and is not implied by the diagram alone: our sun is most likely younger than most stars in the bulge, not older.

GED Practice

Identify Implications

Directions: Choose the one best answer to each question.

Questions 1 and 2 refer to the following passage and diagram.

The diagram below shows the process of photosynthesis, by which plants make their own food using the energy from sunlight.

Photosynthesis

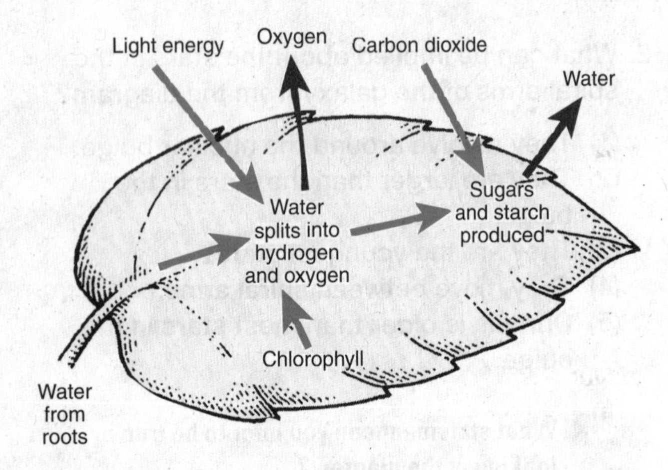

1. According to the diagram, how does chlorophyll differ from the other ingredients of photosynthesis?
   (1) It is broken down into hydrogen and oxygen.
   (2) It combines with oxygen to form sugars and starch.
   (3) It is built from carbon and oxygen.
   (4) It originates inside the plant.
   (5) It is released into the atmosphere.

2. In the diagram, it is implied that what two things are results released during photosynthesis?
   (1) water and oxygen
   (2) hydrogen and oxygen
   (3) chlorophyll and hydrogen
   (4) carbon and chlorophyll
   (5) glucose and hydrogen

Question 3 refers to the following passage.

Exposure to certain kinds of electromagnetic waves can cause health problems. For example, X-rays can cause radiation burns. Very large amounts of microwaves can overheat human cells. In animals, extensive exposure to microwaves has caused damage to the nervous and reproductive systems.

3. Which of the following statements is implied by the passage?
   (1) Exposure to very large amounts of microwaves should be avoided.
   (2) Humans are not at risk from exposure to microwaves.
   (3) Microwaves generated by microwave ovens are not dangerous.
   (4) X-rays can cause skin cancer.
   (5) Health problems related to the nervous system can be prevented by avoiding exposure to microwaves.

TIP

To determine if a fact or idea is implied by a passage, ask yourself if the idea is suggested by something specific in the passage. If it is, the idea or fact is an implication because it can be inferred logically.

Questions 4 through 7 refer to the following diagram.

Life Cycle of a Medium-Sized Star

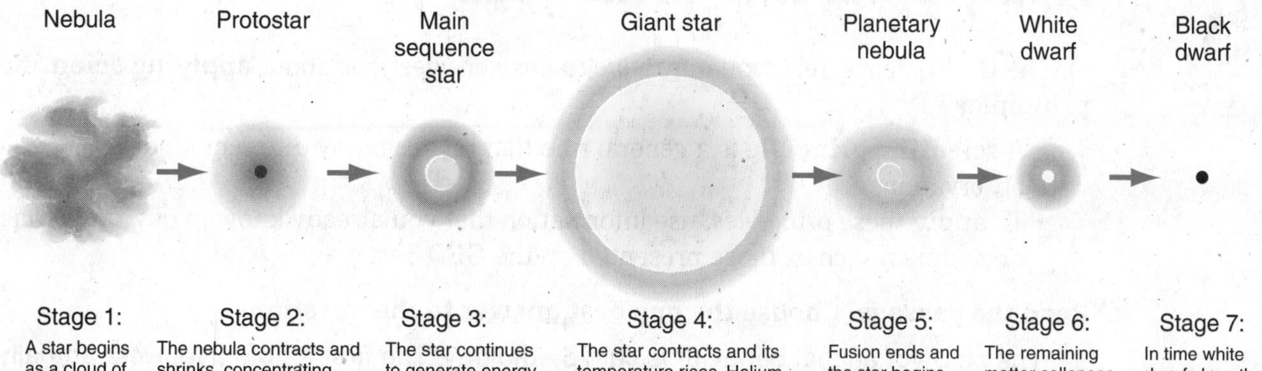

Stage 1:	Stage 2:	Stage 3:	Stage 4:	Stage 5:	Stage 6:	Stage 7:
A star begins as a cloud of dust and gas.	The nebula contracts and shrinks, concentrating matter in a central region. Nuclear fusion begins.	The star continues to generate energy through fusion, using hydrogen as a fuel.	The star contracts and its temperature rises. Helium fusion begins, and the star expands to a size at least 10 times as big as our sun.	Fusion ends and the star begins losing its outer gases.	The remaining matter collapses inward and forms a hot, dense core which gradually cools.	In time white dwarfs lose their heat and sit in space as cold, dark masses.

4. Which of the following sentences is implied by the diagram?
   (1) All stars go through these seven stages.
   (2) Some main sequence stars do not become giant stars.
   (3) Our sun is currently a giant star.
   (4) Fusion occurs during all stages of a star's life.
   (5) The life cycle of a medium-sized star is fairly predictable.

5. Which of the following statements is suggested by the diagram?
   (1) Each stage in a star's life takes about the same amount of time.
   (2) A star reaches its largest size during Stage 4.
   (3) Some stars do not begin as nebulas.
   (4) Stage 2 is the quickest stage of a star's life.
   (5) A white dwarf turns into a planetary nebula.

6. Which of the following happens only during Stages 2, 3, and 4?
   (1) expansion
   (2) contraction
   (3) fusion
   (4) explosion
   (5) cooling

7. Our sun, a medium-sized star, has been a main sequence star for about 5 billion years. What is suggested by this fact and the diagram?
   (1) Our sun's life cycle is almost over.
   (2) Our sun will become a giant star in the next one million years.
   (3) Our sun will increase in size by at least ten times.
   (4) The life cycle of our sun is not very well understood.
   (5) Scientists know the date that our sun will become a giant star.

Answers and explanations start on page 106.

KEY Skill 5

Application

Apply Scientific Principles

On the GED Science Test, you may have to answer questions about **applying scientific principles.**

- A **scientific principle** is a general rule that relates many different kinds of observations.
- To **apply** these principles, use information that you already know in new situations or contexts, such as those presented on the GED Test.

Read the passage. Choose the one best answer to the question.

The study of animal behavior is called ethology. The first ethologists were actually naturalists who observed and described the habits of living animals, often in great detail. In the 1950s a German ethologist, Konrad Lorenz, helped make ethology more systematic with a more scientific approach.

First, Lorenz showed how early impressions of their parents influence the social behavior of birds, such as ducks and geese. In a series of experiments, Lorenz took newly hatched goslings from their parents and raised them himself. As goslings, the birds followed him around as if he were their mother, and as mature geese, they tried to mate with him.

Second, by observing the courtship rituals of ducks, Lorenz discovered that each species has a unique behavior pattern that is as characteristic of that species as the color of its feathers or the shape of its wings. From this, Lorenz concluded that a species' behavior has evolved by natural selection and is adapted to its environment.

QUESTION: Which of the following is an example of a behavior by a species to adapt to its environment?

(1) Male peacocks have brightly colored tail feathers.
(2) Horses walk on four legs.
(3) The white coat of an arctic fox provides camouflage against snow and ice.
(4) Beavers build a lodge or dam in a stream to provide safe shelter.
(5) Some lizards can change color to match their environment.

EXPLANATIONS

STEP 1 To answer this question, ask yourself:

- What is the passage about? how environment influences animals' behaviors
- What rule do I need to apply to answer this question? animals' behavior is a response to their environment

STEP 2 Evaluate all of the answer choices and choose the best answer.

(1) No. The feathers are a physical characteristic, not a behavior.
(2) No. This is an inherited physical ability, not a behavior.
(3) No. Camouflage is a physical response to the environment and not a behavioral response to the environment.
(4) **Yes. The lodge or dam contributes to the survival of the beaver, so it is an example of a behavior that has evolved through natural selection.**
(5) No. Changing color is a defense mechanism, not a learned behavior.

ANSWER: (4) Beavers build a lodge or dam in a stream to provide safe shelter.

Practice the Skill

Try these examples. Choose the one best answer to each question. Then check your answers and read the explanations.

Questions 1 and 2 refer to the following passage and diagram.

Hammers, screwdrivers, and wedges are all examples of machines. A machine is any device that allows you to do work more easily than doing the work with just your body. The force you exert to do the work is called the effort force. Machines change the effort force needed in three ways. They change the force's size, direction, or speed. The diagram shows one example of a machine—a seesaw.

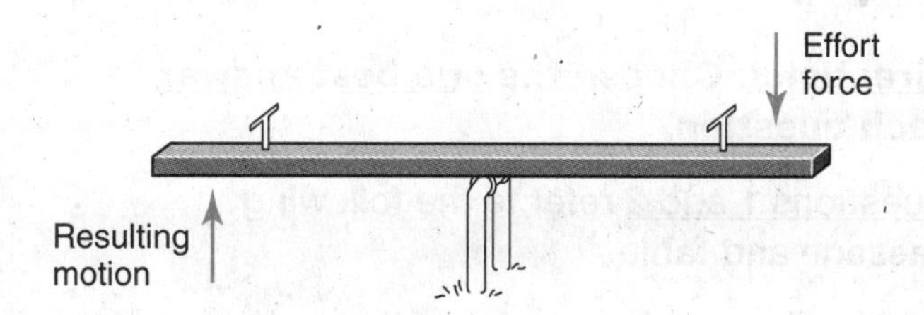

1. Based on the passage and the diagram, how does the seesaw change the effort force?
   (1) It increases the size of the force.
   (2) It decreases the size of the force.
   (3) It changes the force's direction.
   (4) It increases the force's speed.
   (5) It decreases the force's speed.

HINT **Apply the definition of a machine to the diagram.**

2. Which of the following is an example of a machine being used to do work?
   (1) An artist draws with a pencil.
   (2) A painter opens a can with a crowbar.
   (3) A child lifts a rock from the ground.
   (4) A bowler knocks over a bowling pin with a bowling ball.
   (5) A girl pushes a swing with her hand.

HINT **Which of the examples change the force's size, direction, or speed?**

Answers and Explanations

1. (3) It changes the force's direction.
Option (3) is correct because the direction of the resulting motion is opposite the direction of the effort force.

Using a seesaw does not increase the size (option 1) or decrease the size (option 2) of the force. A seesaw does not change the speed of the force (options 4 and 5).

2. (2) A painter opens a can with a crowbar.
Option (2) is correct because a crowbar allows you to change the direction of the force, and in turn, complete the work with less effort. Therefore, a crowbar is a machine.

Options (1) and (4) are incorrect because they involve the use of devices that do not change the force's size, direction, or speed, and therefore are not machines. Options (3) and (5) do not involve machines at all—the workers are performing their tasks manually.

Apply Scientific Principles

Directions: Choose the one best answer to each question.

Questions 1 and 2 refer to the following passage and table.

The diagram below shows three classes of levers, which are among the most commonly used machines for multiplying force. On each lever, the effort point is where the effort force is applied. The work point is the location where work is done. The fulcrum is the location where the lever is supported.

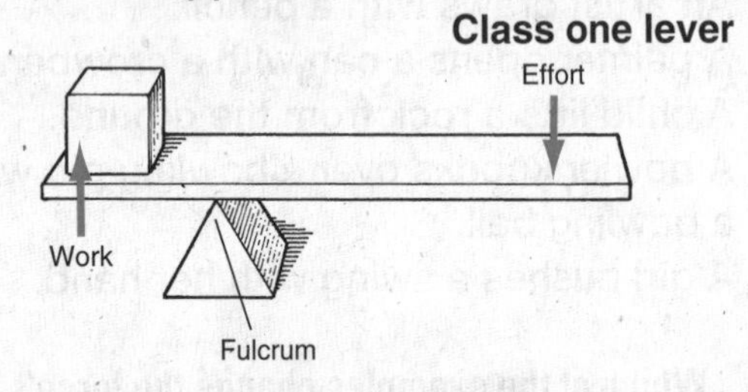

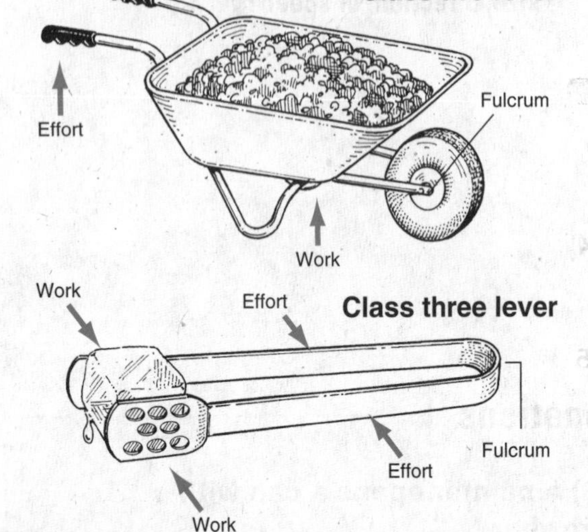

1. Which of the following is an example of a class one lever?
   (1) tweezers, which have effort points between the fulcrum and work points
   (2) a paper cutter, which has a work point between the fulcrum and the effort point
   (3) a hammer, which has no fulcrum
   (4) pruning shears, which have a fulcrum between the effort and work points
   (5) a nutcracker, which has work points between the fulcrum and effort points

2. Which of the following observations is based on the diagram?
   (1) Class one levers do more work than class two levers.
   (2) Only class two and class three levers have fulcrums.
   (3) Classes of levers vary in the relative positions of their fulcrums, work points, and effort points.
   (4) All class two levers have two effort points.
   (5) Class three levers do more work than class one levers.

Question 3 refers to the following passage and graph.

Male warblers sing songs to attract females during mating season. The graph below compares the number of different songs sung by a male to the number of days it takes a male to pair with a female warbler.

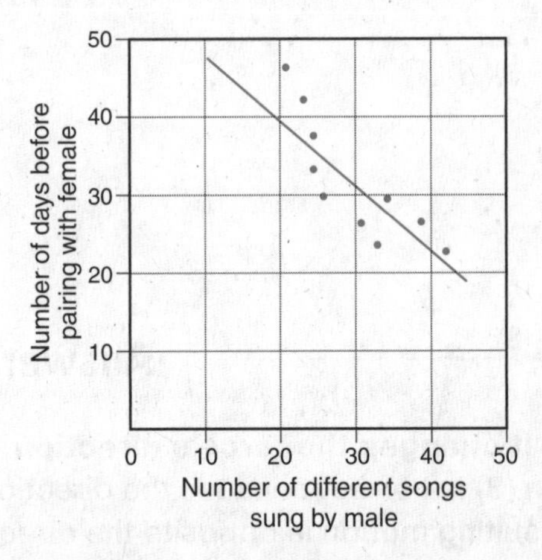

3. What principle would be accurate based on the information given?
   (1) All warblers sing exactly the same songs.
   (2) Only male warblers sing.
   (3) Warblers sing only during mating season.
   (4) A warbler needs to know at least 40 songs in order to find a mate.
   (5) Male warblers that sing more songs tend to find mates sooner than male warblers that sing fewer songs.

<u>Questions 4 through 6</u> refer to this map.

Earth's Major Earthquake Zones

Eurasian–Melanesian Belt
Europe
Asia
Africa
North America
Mid-Atlantic Ridge
Ring of Fire
Pacific Ocean
South America
Australia
Atlantic Ocean
Other Earthquake Areas

4. Based on the map, which of the following statements is most accurate?
   (1) More earthquakes occur each year in Africa than Australia.
   (2) Earthquakes can occur both on land and below the oceans.
   (3) More earthquakes occur along the Mid-Atlantic Ridge than along the Eurasian-Melanesian belt.
   (4) Earthquakes never occur in the Pacific Ocean.
   (5) Earthquakes never occur in the central United States.

5. Tsunamis, or giant waves, can be triggered by underwater earthquakes. Which of the following areas is probably not in danger of being hit by a tsunami?
   (1) the eastern coast of Africa
   (2) the northeast coast of Australia
   (3) the southern coast of Australia
   (4) the southwest coast of North America
   (5) the southeast coast of South America

6. Earthquakes are caused by movements of Earth's plates. Most earthquakes occur along or near the boundaries of large plates. Applying this principle, what can you infer from this information and the map?
   (1) A boundary between plates occurs along the southeast coast of South America.
   (2) The Ring of Fire is one of Earth's major earthquake zones.
   (3) Earthquakes only occur near the continents.
   (4) A boundary between plates exists beneath where belts or strings of earthquakes are shown on the map.
   (5) Earthquakes never occur outside the major earthquake zones.

TIP

To determine whether an answer choice is applying a scientific principle from the passage or diagram, ask yourself: *Does this answer support the information in or the general rule of the passage or diagram?*

Answers and explanations start on page 106.

Analysis

Make Inferences

On the GED Science Test, you may have to **make inferences** based on information presented in a passage or illustration.

- An **inference** is a conclusion based on evidence and information.
- You **make an inference** or **infer** when you determine that something is true based on information that is presented but not specifically stated as a fact.

After you read the passage below, you will make an inference about waves.

Read the passage. Choose the one best answer to the question.

Waves are disturbances that move through the particles of a material or through a vacuum. Waves transfer energy. As a wave moves through a material, energy is transferred from one particle to another. The material that the waves move through is called the medium.

Water, sound, and seismic (earthquake) waves are examples of mechanical waves, which require a medium to carry their energy. X-ray, radio, and light waves are examples of electromagnetic waves, which don't require a medium, so they can travel through a vacuum.

QUESTION: Which of the following statements is an inference that you can make based on the information in the passage?

(1) Seismic waves travel by the transfer of energy from one rock particle to another.
(2) Matter varies in its ability to carry sound waves.
(3) All waves require a medium to carry energy.
(4) Waves transfer energy.
(5) Waves are disturbances that move through a vacuum.

EXPLANATIONS

STEP 1 To answer this question, ask yourself:

- What does the passage say about waves? how the two types of waves travel and what they travel through
- What is the question asking me to do? use common sense and the information given to infer, or figure out, something that is not explicitly stated in the passage

STEP 2 Evaluate all of the answer choices and choose the best answer.

(1) **Yes. Because seismic waves are mechanical waves, you can infer that they move through a medium—particles in the rock.**
(2) No. There is no information in the passage to support this inference.
(3) No. This contradicts information in the passage about electromagnetic waves.
(4) No. This is a direct statement from the passage, not an inference based on evidence.
(5) No. This contradicts information in the passage about mechanical waves.

ANSWER: (1) Seismic waves travel by the transfer of energy from one rock particle to another.

Practice the Skill

Try these examples. Choose the one best answer to each question. Then check your answers and read the explanations.

Questions 1 and 2 refer to this passage and graph.

According to the big bang theory, the universe has been expanding since it began 15 billion years ago. This expansion is countered by gravity, an attractive force that each object in the universe exerts on all other objects.

If there is enough matter in the universe, the force of gravity will eventually stop the expansion and the universe will begin contracting. If there is not enough matter for gravity to stop the expansion, the universe will continue expanding forever. If there is just the right amount of matter, the expansion will keep getting slower and slower, although it will never actually stop.

Possible Fates of the Universe

Open universe
Flat universe
Closed universe
Big crunch
Size of the universe
Time
Big bang

1. What inference can you make about the open universe scenario?

(1) It is based on there being enough mass for gravity to collapse the universe.
(2) It will end in a "big crunch."
(3) The universe will continue expanding forever.
(4) It is based on a force that pushes the universe apart.
(5) It began 15 billion years ago.

HINT **What information in the text can you use to understand the graph?**

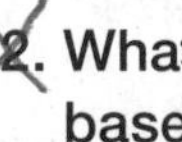

2. What can you infer about the force of gravity based on the passage?

(1) The force of gravity increases as matter increases.
(2) Gravity did not exist 15 billion years ago.
(3) In the open universe, gravity eventually stops working.
(4) Because gravity exists, only the closed universe could actually exist.
(5) Gravity is a repellant force that causes expansion.

HINT **Read the passage carefully for information about how gravity applies to each scenario.**

Answers and Explanations

1. (3) The universe will continue expanding forever.

Option (3) is correct because the graph of the open universe matches the description of a universe that expands forever.

The graph does not show a collapse (option 1) or big crunch (option 2) for the open universe. There is no force mentioned that pushes the universe apart (option 4). The age of the universe (option 5) is stated directly in the passage, not inferred.

2. (1) The force of gravity increases as matter increases.

Option (1) is correct because the passage indicates that the effect of gravity on the universe increases as the amount of matter increases.

There is no mention in the passage of gravity not existing at the big bang (option 2) or in the future (option 3), and there aren't enough facts to make these inferences. The statement that only a closed universe can exist (option 4) contradicts the passage. Option (5) confuses facts—gravity is an attractive force that acts against expansion.

GED Practice

Make Inferences

Directions: Choose the one best answer to each question.

Questions 1 and 2 refer to this passage and diagram.

A leaf's structure, with layers of tissue cells, supports its role as a center for photosynthesis. The cuticle consists of a single layer of epidermal cells that secrete a waxy substance. It is almost waterproof, so even on hot days a leaf rarely dries out completely. The cuticle is also transparent, so light passes through it to the second layer of tissue.

The second layer, the palisade layer, consists of cells that contain most of the plant's chlorophyll. Photosynthesis takes place in the palisade layer.

The sponge cells of the third layer soak up most of the water that reaches the leaf. Spaces between water-logged sponge cells are filled with water vapor and carbon dioxide. These substances are needed for photosynthesis.

Like the cuticle, the guard cells are waxy and transparent. However, the layer of guard cell is filled with pores called stomata. The stomata allow carbon dioxide to enter the leaf and oxygen to leave. By opening and closing, the stomata also control the rate at which water evaporates from the leaf.

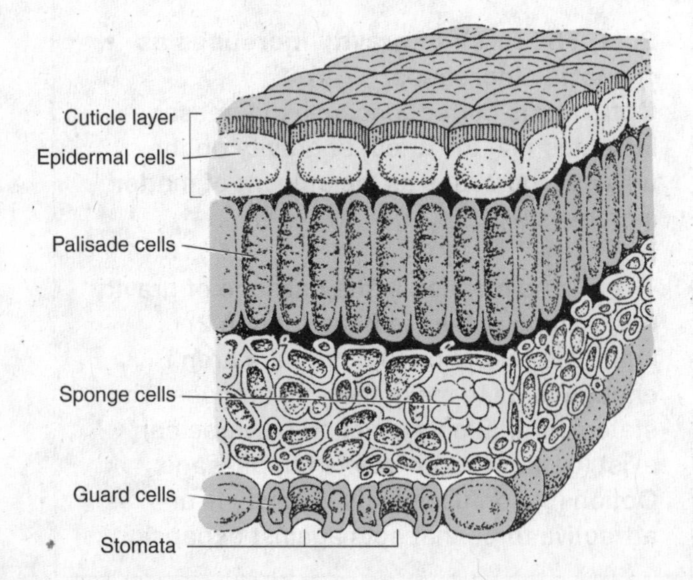

1. What can you infer about how palisade cells and sponge cells interact in a leaf?
   (1) Photosynthesis takes place in both types of cells.
   (2) They are in different layers, so they do not interact at all.
   (3) Sponge cells protect palisade cells.
   (4) Sponge cells store the materials that palisade cells use during photosynthesis.
   (5) The palisade cells and the sponge cells protect the epidermal cells.

2. Which human organ performs a function most similar to that of a leaf's cuticle?
   (1) eyes
   (2) lungs
   (3) stomach
   (4) skin
   (5) bones

Question 3 refers to this passage.

When water evaporates, it absorbs energy in the form of heat. The added energy causes the water molecules to move faster. If they gain enough energy, their energy of motion exceeds the forces that hold them together in the liquid. These molecules then become water vapor, the gas phase form of water.

3. Which of the following is not directly stated but can be inferred from the passage?
   (1) Molecules in liquid water are in motion.
   (2) Water can evaporate only at 100°C.
   (3) Water in its liquid phase is called water vapor.
   (4) During evaporation, water molecules release energy into the environment.
   (5) Water is an essential material for all living things.

Questions 4 and 5 refer to this passage and diagram.

People have studied the patterns of stars, called constellations, for thousands of years. Different cultures have seen different representations in the arrangement of stars, and they have named many of them. The stars in a constellation appear fairly close together in the sky.

Many of the constellations we use to map the night sky are named after ancient gods, heroes, and animals. The constellation Orion, named for Orion the Hunter, can be seen in winter months in the northern hemisphere. Rigel is the brightest star in Orion. Stars labeled in the diagram represent Orion's right shoulder, belt, and left foot.

Orion the Hunter

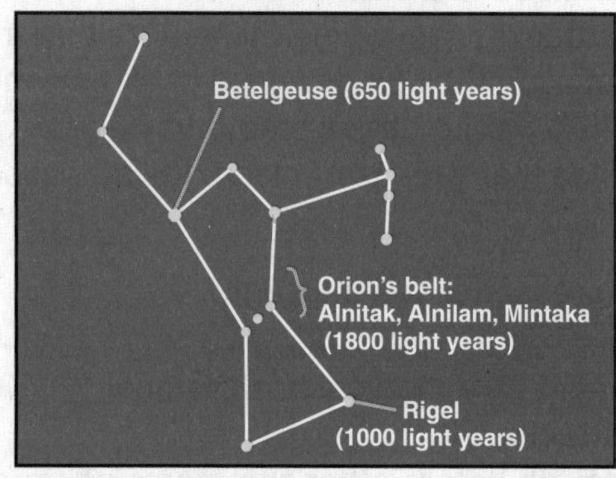

4. Which of the following statements can you infer from the passage or diagram?
   (1) Stars that are closest to Earth appear brightest.
   (2) The stars in Orion's belt are older than the stars Rigel and Betelgeuse.
   (3) Different constellations are visible during different seasons.
   (4) All of the stars in Orion are the same distance from Earth.
   (5) Stars do not move in the sky.

5. What can you infer about constellations from the diagram?
   (1) All stars are between 650 and 1800 light years from Earth.
   (2) The stars in Orion's belt are dim compared to the stars in other constellations.
   (3) All of the stars in Orion and other constellations are smaller than our sun.
   (4) The sky contains few constellations.
   (5) As seen from Earth, stars appear close together, but they are actually light years apart.

Question 6 refers to this passage.

Much of the very hot matter in our universe emits radiation as X-rays, which can be observed by X-ray telescopes. In July of 1999, NASA launched the Chandra X-ray Observatory. Chandra can capture images 25 times sharper than previous X-ray telescopes. Among its many observations, Chandra has collected X-rays that provide data about quasars that are 10 billion light years away.

6. Which of the following is suggested by the passage?
   (1) Scientists no longer need older telescopes.
   (2) NASA will launch a new telescope soon.
   (3) X-ray telescopes will never be able to see farther than 10 billion light years away.
   (4) Chandra helps scientists learn more about the hot matter in our universe.
   (5) Chandra was launched in 1999.

TIP

An inference frequently connects two separate facts. As you read passages or diagrams, try to find the connections between facts.

Answers and explanations start on page 107.

Analysis

Identify Facts and Opinions

On the GED Science Test, you may have to answer questions in which you will distinguish between **fact** and **opinion.**

- A **fact** is a piece of information that can be proven as true.
- An **opinion** is someone's point of view about a topic, which may or may not be true.

As you read the passage below, distinguish between the facts and opinions about asthma.

Read the passage. Choose the one best answer to the question.

Many children have asthma, an illness in which the small air passages in the lungs contract, causing breathing difficulties. The symptoms of an asthma attack include wheezing, more time spent exhaling than inhaling, and faster breathing. Some asthma attacks are so serious that they require medical attention.

Asthma attacks can be caused by colds or bronchitis, exercise, air pollution, or an allergic reaction to something such as dust, pollen, or animal hair. Some doctors think that emotional stress can also cause an asthma attack.

Treatment of asthma involves preventing attacks by avoiding the factors that trigger them. Patients can also monitor the amount of air in their lungs with a peak flow meter. Lung capacity can decrease by as much as 25% before other symptoms appear. Asthma attacks can be controlled with medicine that widens or clears the air passages in the lungs.

QUESTION: Which statement is an opinion expressed in the passage?

(1) Asthma only affects children.
(2) Some asthma episodes require treatment by a medical professional.
(3) Emotional stress can cause asthma attacks.
(4) Lung capacity can decrease by one-fourth before other symptoms appear.
(5) Prescription drugs can be harmful to children.

EXPLANATIONS

STEP 1 To answer this question, ask yourself:

- What information did the passage provide? causes, symptoms, and treatment of asthma
- How can I answer the question? determine what information represents a point of view and is not based on factual information

STEP 2 Evaluate all of the answer choices and choose the best answer.

(1) No. While the passage is about asthma in children, there is no indication that it does not affect adults.
(2) No. This is a fact presented in the passage.
(3) **Yes. The phrase "Some doctors think..." indicates that this is an opinion rather than a fact.**
(4) No. This is a fact presented in the passage.
(5) No. This is an opinion, but prescription drugs are not presented in the passage.

ANSWER: (3) Emotional stress can cause asthma attacks.

Practice the Skill

Try these examples. Choose the one best answer to each question. Then check your answers and read the explanations.

Questions 1 and 2 refer to the following passage and graph.

Many studies have shown a link between smoking and lung cancer deaths. Although smoking is not the only cause of lung cancer, as the number of cigarettes smoked increases, so do the chances that a person will contract lung cancer.

All smokers should quit smoking. Quitting can reduce a smoker's risk of lung cancer. However, an ex-smoker still has some risk of the disease. A person who has never been a smoker has the lowest risk.

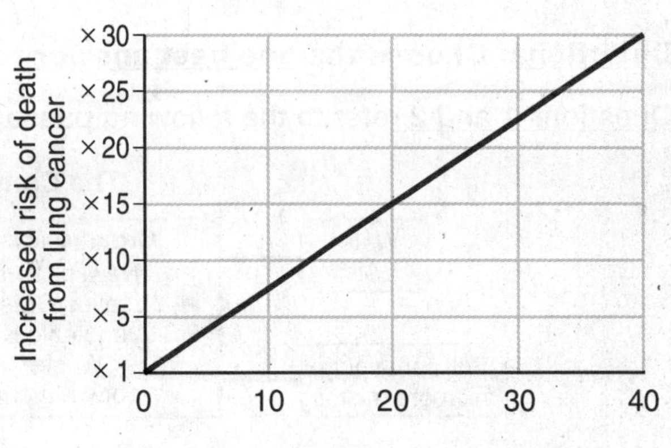

1. Which sentence is an opinion supported by the graph and passage?
   (1) Many studies have shown a link between smoking and lung cancer.
   (2) Smoking is not the only cause of lung cancer.
   (3) Ex-smokers are still at risk for lung cancer.
   (4) All smokers should quit smoking.
   (5) Non-smokers have the lowest risk for lung cancer.

HINT **Which choice expresses a point of view supported by the passage?**

2. Which of these statements is a fact according to the passage and graph?
   (1) Lung cancer rates have decreased over the past 25 years.
   (2) Smoking may be considered one of the greatest health risks.
   (3) Lung cancer rates increase as the number of cigarettes smoked increases.
   (4) The leading cause of death among smokers is lung cancer.
   (5) People should stop smoking cigarettes.

HINT **Eliminate wrong answer choices by looking for words that indicate the statement is not a fact.**

Answers and Explanations

1. (4) All smokers should quit smoking.
Option (4) is correct. This statement is an opinion that some people may disagree with. It cannot be proven true.

Options (1), (2), (3), and (5) are incorrect. These statements are all facts that can be proven true with medical research and historical examples.

2. (3) Lung cancer rates increase as the number of cigarettes smoked increases.
Option (3) is correct because there is a direct correlation between the number of cigarettes smoked and lung cancer deaths.

There is no data presented about changes in lung cancer deaths over time (option 1) or about other causes of death among smokers (option 4). The phrase "may be considered" indicates that the statement is an opinion, not a fact (option 2). The statement about what people "should" do (option 5) is an opinion supported by the facts.

GED Practice

Identify Facts and Opinions

Directions: Choose the one best answer to each question.

Questions 1 and 2 refer to the following passage and diagram.

The Composting Process

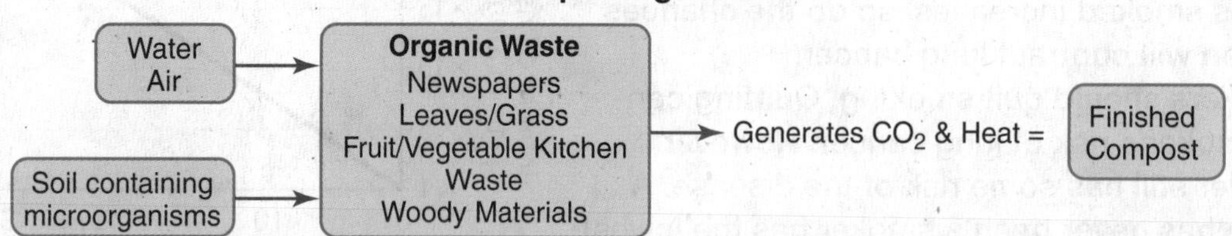

Composting is one way to produce rich, organic material that can be used to enhance soil. Many people prefer compost made from kitchen waste rather than from lawn waste.

During composting, microorganisms such as bacteria and fungi from soil consume the organic waste and break it down into smaller particles. These microorganisms require water and air to live and multiply. As they break down the organic waste, the bacteria and fungi give off carbon dioxide and heat that speed up the process of decay, transforming organic waste into rich compost. Because of this transformation, some people think all organic waste should be composted.

The finished compost acts as an organic fertilizer, which some gardeners believe is better than synthetic fertilizers, and provides essential nutrients for healthy plants.

1. Which of the following statements is an opinion about composting?
   (1) Compost can be used to enhance soil.
   (2) Compost is better for the garden than synthetic fertilizers.
   (3) Composting cannot occur without microorganisms from soil.
   (4) Heat is produced during composting.
   (5) A supply of carbon dioxide is necessary for composting to occur.

2. Which of these statements is a fact about composting?
   (1) Compost made with kitchen waste is better than compost made with lawn waste.
   (2) Composting generates carbon dioxide and heat.
   (3) Compost is better than synthetic fertilizers.
   (4) Adding a small amount of iron to the mixture will improve the compost.
   (5) All organic waste should be composted.

Question 3 refers to this passage.

During fission, the nucleus of a heavy atom, such as uranium, is split apart and releases a large amount of energy. There are several problems with fission energy. Uranium is a nonrenewable resource and will eventually be used up. Nuclear power plants are expensive to build, and there is the chance that harmful radiation could escape from a plant in an accident. Many people think that nuclear power plants should not be located near cities because of this risk.

3. Which statement is an opinion?
   (1) Building nuclear power plants is expensive.
   (2) Energy is released when atoms split.
   (3) Nuclear power is renewable energy.
   (4) Nuclear fission is currently being used to produce power.
   (5) Nuclear power plants are too dangerous to be located near cities.

Questions 4 and 5 refer to the following passage and bar graph.

Fossil fuels such as gas, oil, and coal are nonrenewable resources because it takes millions of years to replace them. One study made predictions about how long the remaining reserves of fossil fuels will last at current usage rates. The results are displayed on the graph. Some scientists think that there are undiscovered deposits of oil and gas that could extend past the predicted dates on the graph. However, if current use continues, all reserves will eventually be consumed. Alternative sources are being researched.

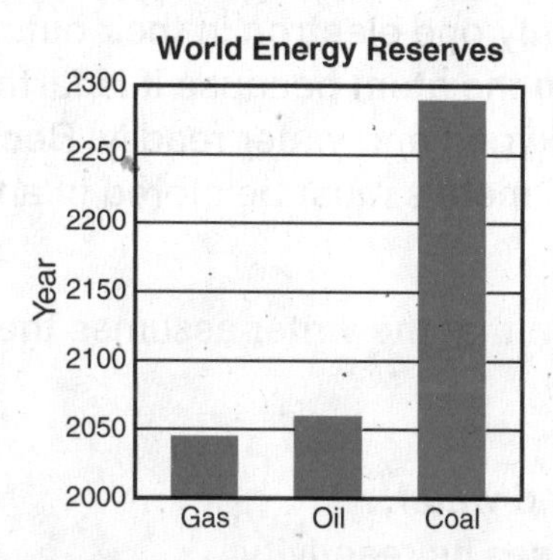

4. Which of the following statements is most likely to be an opinion?

(1) Known reserves of coal will last longer than those of oil if current use continues.
(2) Oil and gas are nonrenewable resources.
(3) There are large undiscovered reserves of fossil fuels.
(4) Coal is a fossil fuel.
(5) Some energy comes from alternative sources.

5. Which statement is not an opinion about the information in the diagram and passage?

(1) Coal is a better fuel than oil.
(2) Oil companies should find new fuel reserves that are currently unknown.
(3) Gas and oil should not be burned for energy because they are valuable fossil fuels.
(4) The energy gained from coal is not enough to justify the environmental damage due to mining.
(5) It takes millions of years to replace fossil fuels.

Question 6 refers to the following passage.

Cloud seeding is the attempt to change the amount or type of precipitation that falls from clouds. It is done by spraying substances into the air that serve as nuclei for the formation of ice particles. The most common chemicals used for cloud seeding include silver iodide and dry ice (frozen carbon dioxide).

A substance such as silver iodide, which has a crystalline structure similar to that of ice, can cause water vapor to freeze. If there is enough growth of ice crystals, the particles become heavy enough to fall from clouds that otherwise would produce no precipitation.

Cloud seeding success is difficult to track. Rain or snow is often observed after the clouds are seeded. However, it is not possible to be certain that the precipitation would not have formed without the seeding. Even if it does work, some scientists believe that cloud seeding can create problems for other areas. Rain that would have formed in places downwind of the seeding cannot occur because the water has been removed.

6. Which of the following statements is an opinion according to the passage?

(1) It is hard to track cloud seeding.
(2) Cloud seeding can create problems for places that are downwind of the seeding.
(3) Silver iodide has a crystal structure that is similar to that of ice.
(4) The most common chemicals used for cloud seeding are silver iodide and dry ice.
(5) Rain is always observed after cloud seeding.

TIP

When trying to determine whether a statement is fact or opinion, look for key words such as *believe, may, think, should,* and *suggest*. These words are most often associated with opinions. Facts will usually be located within information in the passage or diagram.

Answers and explanations start on page 107.

Skill 8

Analysis

Recognize Assumptions

On the GED Science Test, you may have to answer questions in which you identify assumptions made in a passage or diagram.

- An **assumption** is knowledge that the writer or illustrator takes for granted and assumes the audience already knows.
- To **recognize assumptions,** ask what you need to already know in order to understand the information that is presented.

Read the passage. Choose the one best answer to the question.

The alkali metals—which include lithium, sodium, and potassium—are considered the most metallic of the elements. In their pure forms, each of these elements is a light, soft, and shiny metal. Alkali metals have only one electron in their outer energy level. This single electron is easily removed from the atom because it is farthest away from the nucleus. The alkali metals react with oxygen and water readily. Because of the energy released by these reactions, alkali metals must be stored in an environment with absolutely no oxygen.

QUESTION: Which statement contains information that the writer assumes the reader already knows?

(1) The majority of elements are metallic.
(2) Alkali metals readily react with oxygen and water.
(3) Removing electrons from an atom increases its reactivity.
(4) Calcium is an alkali metal.
(5) When an alkali metal loses an electron, a positive ion forms.

EXPLANATIONS

STEP 1 To answer this question, ask yourself:

- What does this passage describe? properties and reactivity of alkali metals
- What information do you need to know that is not stated? Electrons are involved in reactions of atoms.

STEP 2 Evaluate all of the answer choices and choose the best answer.

(1) No. The reader does not need to know how many elements are metals in order to understand the passage.
(2) No. This information is stated in the passage, so it is not an assumption.
(3) **Yes. The phrases "...because it is farthest away from the nucleus," and "The alkali metals react...readily," indicate the writer assumes the reader already knows that electrons affect the reactivity of an element.**
(4) No. The reader does not need to have knowledge about calcium, and calcium is not an alkali metal.
(5) No. Though true, this information is not necessary in order to understand the concepts of the passage.

ANSWER: (3) Removing electrons from an atom increases its reactivity.

Practice the Skill

Try these examples. Choose the one best answer to each question. Then check your answers and read the explanations.

Questions 1 and 2 refer to this passage and diagram.

All organisms use energy in their life functions. The diagram below shows an example of a food chain in which energy passes from one organism to another. Each element of the food chain receives energy from the previous element.

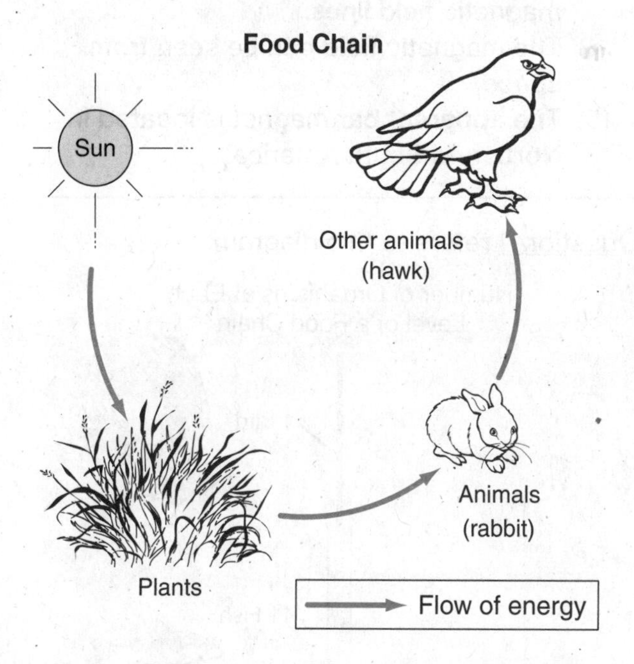

1. What is assumed to be the original source of energy in a food chain?

   (1) animals, which use energy for their life functions
   (2) plants, which get energy from other plants
   (3) light, which is energy from the sun
   (4) the ocean, because many tiny organisms live there
   (5) chemicals, which are obtained from digested food

HINT **Where does the food chain start?**

2. What information does the diagram assume the reader knows?

   (1) All organisms use the sun's energy to make food.
   (2) Rabbits prefer to eat grass more than seeds.
   (3) Further away from the sun in the food chain, the amount of energy decreases.
   (4) Hawks are before rabbits on the food chain.
   (5) Energy is transferred to animals when they eat something that comes before them in the food chain.

HINT **Which statement explains something that is not shown directly?**

Answers and Explanations

1. (3) light, which is energy from the sun
Option (3) is correct because it can be assumed that the energy from the sun needed for photosynthesis is in the form of light. This light is the original source of energy according to the diagram.

Animals (option 1) are not shown until the third step of the diagram, so they are not the original source. Plants do not obtain energy from other plants (option 2), so that would be a poor assumption. The ocean (option 4) is not mentioned in the diagram or passage; neither are chemicals from digested food (option 5).

2. (5) Energy is transferred to animals when they eat something that comes before them in the food chain.
Option (5) is correct because the diagram assumes the reader knows that animals in the food chain eat plants or other animals, and that is the mechanism for transferring energy.

Most animals can't use the sun's energy directly (option 1). Whether rabbits prefer grass more than seeds (option 2) isn't addressed by the diagram. There is no information supporting option (3) in the passage or diagram. The diagram shows directly that hawks are after rabbits on the food chain, so option (4) contradicts the diagram.

GED Practice

Recognize Assumptions

Directions: Choose the one best answer to each question

Questions 1 to 3 refer to this passage and diagram.

Earth behaves as if a huge bar magnet were buried deep below its surface, as shown in the diagram. The red lines represent a magnetic field, the area where the magnetic force is felt. The poles, at the top and bottom, are where the magnetic force is strongest.

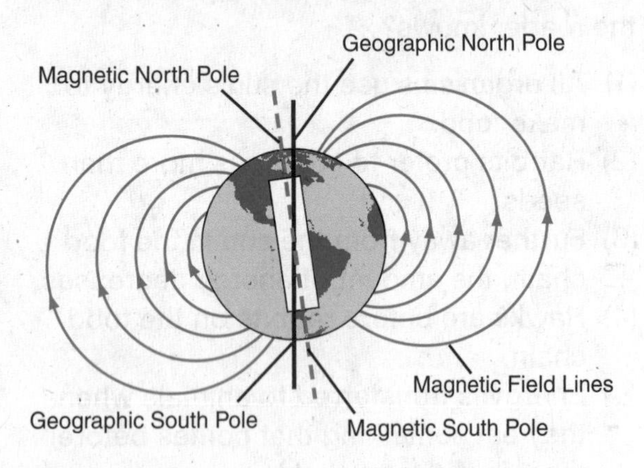

1. Based on the passage, it is assumed that the reader already knows which of the following?
   (1) how to use a magnetic compass
   (2) what magnets are
   (3) how Earth's magnetic poles formed
   (4) that some rocks are natural magnets
   (5) what a magnetic pole is

2. What does the illustrator assume?
   (1) The reader knows how to calculate magnetic field strength.
   (2) The reader is familiar with the term "magnetic field."
   (3) The reader has experimented with magnets.
   (4) The reader has seen Earth's magnetic field.
   (5) Earth's magnetic and geographic poles are in the same location.

3. Which of the following does the illustrator assume that the reader knows?
   (1) Earth's magnetic field extends into space.
   (2) A magnet has two poles.
   (3) Magnetic force is represented by magnetic field lines.
   (4) The magnetic field can be seen from space.
   (5) The apparent bar magnet is located in North and South America.

Question 4 refers to this diagram.

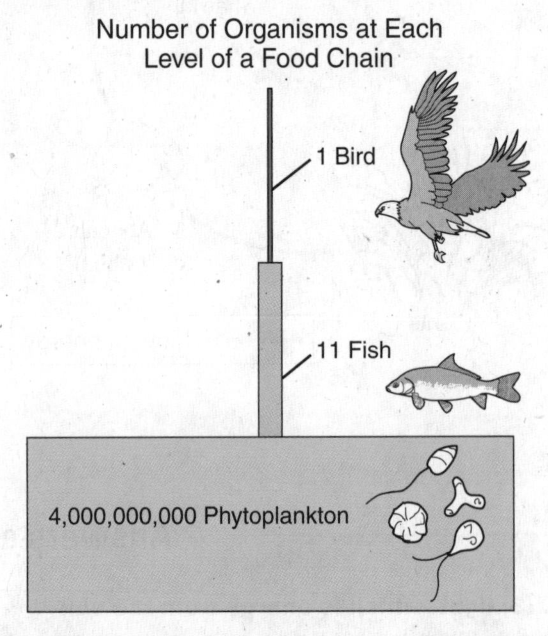

4. The diagram shows how the number of organisms changes with each level in a food chain. What information below does the illustrator assume the reader already knows?
   (1) Birds eat fish.
   (2) Organisms at the top of the food chain are larger.
   (3) All organisms can use sunlight to make food.
   (4) There are more organisms at the top of the chain than at the bottom.
   (5) The animals at the top of the food chain are mammals or birds.

Questions 5 and 6 refer to this figure.

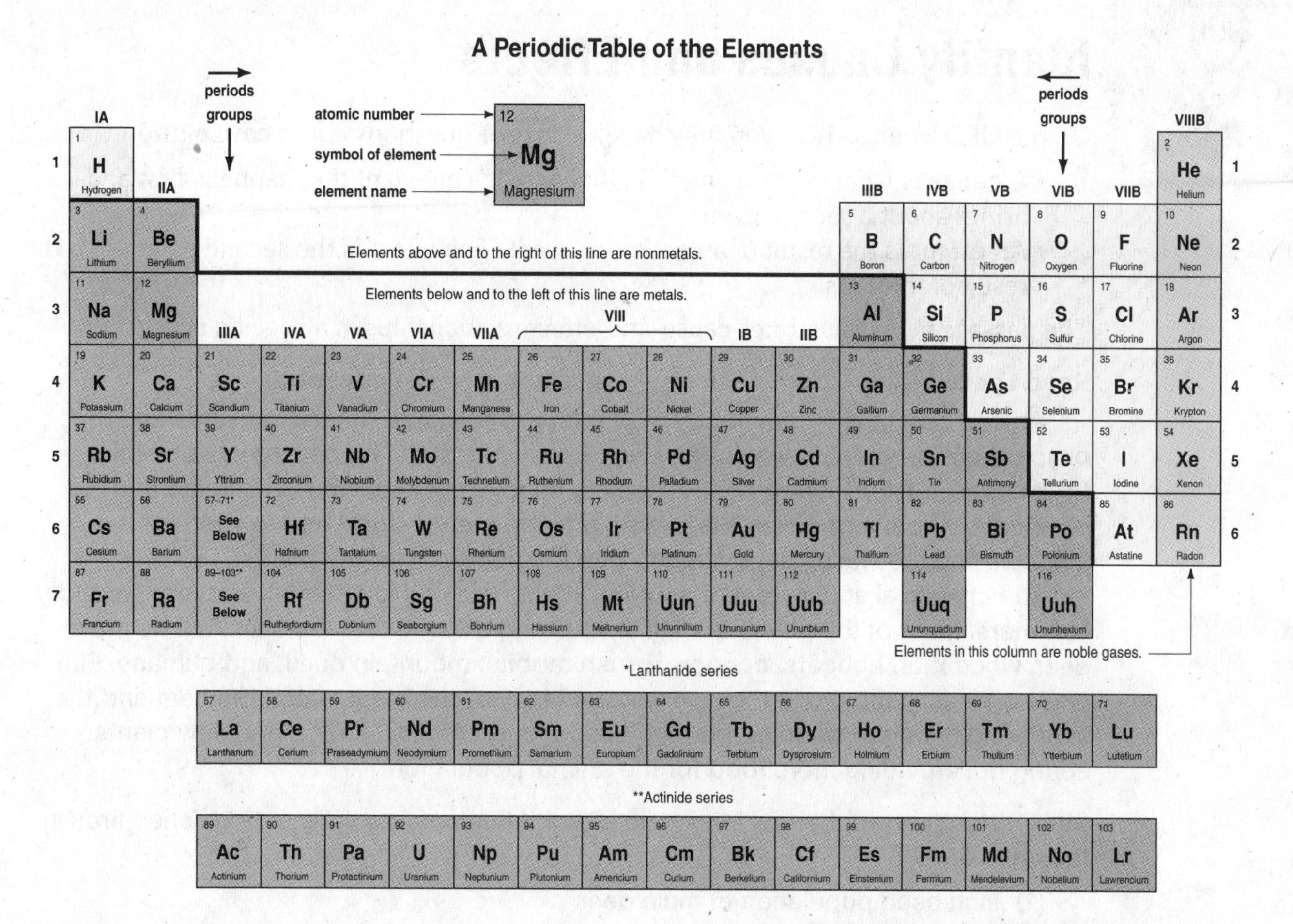

5. Based on the periodic table, it is assumed that the reader already knows which of the following?
   (1) what magnesium looks like
   (2) what part of the periodic table represents a period
   (3) how the atomic numbers of calcium and sodium differ
   (4) which elements are noble gases
   (5) how to read Roman numerals

6. The creator of the periodic table most likely assumed which of the following?
   (1) The reader knows the names of elements in the Actinide series.
   (2) Some elements are radioactive.
   (3) The reader has studied the noble gases.
   (4) The reader knows what atomic numbers represent.
   (5) All elements in the table are solids at room temperature.

TIP

When answering a question about a diagram, be sure to read the title, caption, labels, and keys before answering the question. Ask yourself: *What does the writer or the illustrator assume I know or assume to be true?*

Answers and explanations start on page 108.

Analysis

Identify Causes and Effects

On the GED Science Test, you may have to answer questions about **cause and effect.**

- A **cause** is what makes something happen. It's an event that happens first and brings about a second event.
- An **effect** is the result of an action or event. The effect is the second event—it's the result of the cause.

The passage below illustrates cause-and-effect relationships in a specific ecosystem.

Read the passage. Choose the one best answer to the question.

An ecosystem is a community of plants, animals, and other organisms and the physical environment in which they live. Ecosystems can be as large as an entire forest or as small as a tide pool. A healthy ecosystem is balanced. Climate and physical environment are stable, plants provide enough food and oxygen for the animals, and the waste is recycled to provide nutrients for the plants.

The chaparral ecosystem of southern California has low rainfall and long, hot, dry summers. Most of the plants are tall shrubs and dwarf trees. Animals include mule deer, wood rats, bobcats, cougars, brush rabbits, mountain quail, and humans. Fire plays an important role in this ecosystem. Many of the plants need the heat and the scattering action of wildfires for their seeds to germinate. After a fire, new plants spring up, providing more food for the animal population.

QUESTION: What would be a long-term effect if humans regularly extinguished fires in the chaparral?

(1) increased population of mule deer
(2) decreased population of humans
(3) increased population of plants
(4) decreased population of plants
(5) no significant change

EXPLANATIONS

STEP 1 To answer this question, ask yourself:

- What is this passage about? how the ecosystem of the chaparral stays in balance
- What is the question asking me to do? determine the results if humans regularly put out fires

STEP 2 Evaluate all of the answer choices and choose the best answer.

(1) No. Reduced wildfires will cause fewer plants to grow, providing less food for all animals, including mule deer.
(2) No. Humans are not dependent on the chaparral for food or survival.
(3) No. Plants need the heat and scattering of wildfires to cause seed germination.
(4) **Yes. Putting out fires regularly would disrupt the process that causes plants in the chaparral to germinate.**
(5) No. Humans regularly extinguishing fires will cause change in the ecosystem, specifically a decrease in plants.

ANSWER: (4) decreased population of plants

Practice the Skill

Try these examples. Choose the one best answer to each question. Then check your answers and read the explanations.

When air cools, its molecules lose energy. As a result, water molecules in the air condense, or change from gas to liquid in the form of tiny droplets. Eventually, many tiny droplets of water form fog or clouds. Fog and clouds are masses of water particles visible in the air. Fog forms near Earth's surface and clouds form higher in the atmosphere.

When water droplets in a cloud grow too heavy to remain suspended in the atmosphere, precipitation occurs. Precipitation is any form of water falling from the air to Earth's surface.

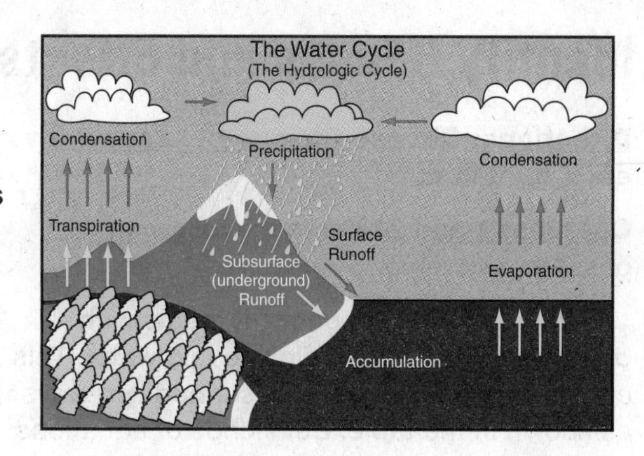

1. What is the effect when water droplets in a cloud grow too heavy to remain suspended in the atmosphere?

   (1) Fog is formed.
   (2) The water cycle is formed.
   (3) Liquid water changes into a gas.
   (4) The air cools.
   (5) Precipitation occurs.

HINT **What happens when the atmosphere cannot hold water droplets?**

2. According to the passage and the diagram, which of the following events causes clouds to form?

   (1) precipitation from clouds
   (2) condensation of water in air
   (3) evaporation from water
   (4) surface runoff from mountains
   (5) transpiration from plants

HINT **Read the passage carefully for information about clouds. Then find information in the diagram to support it.**

Answers and Explanations

1. (5) Precipitation occurs.
Option (5) is correct because according to the passage, when water droplets in clouds get too heavy to be suspended in the air (cause), precipitation occurs (effect).

Fog is formed (option 1) when water molecules condense in the air. Option (2) is incorrect; precipitation is just one small part of the water cycle. Water changing to gas (option 3) occurs in another part of the water cycle, so it can't be an effect of clouds becoming too heavy. The air cooling (option 4) causes clouds and fog to form, and is not an effect of water droplets becoming too heavy.

2. (2) condensation of water in air
Option (2) is correct because when water molecules in air condense, visible water droplets form. Many droplets form a cloud. This is also shown in the diagram by the arrows pointing up and the label *Condensation* under the clouds.

Precipitation (option 1) is an event that occurs after the clouds are already formed. Evaporation (option 3) and transpiration (option 5) are shown in the diagram as events that occur before the condensation of water into clouds. Surface runoff (option 4) is shown in the diagram but not as part of cloud formation.

GED Practice

Identify Causes and Effects

Directions: Choose the one best answer to each question.

Questions 1 and 2 refer to the following passage and table.

When natural gas burns, it reacts with oxygen to form carbon dioxide and water. This chemical reaction can be described two ways, as shown in the table. Both kinds of equations tell you what elements or compounds enter into the chemical change and what elements or compounds are produced.

However, only the chemical equation reveals the amounts of the different substances involved. The chemical equation tells you that, for each molecule of methane (CH_4) that burns, two molecules of oxygen (O_2) react. The result is the formation of one molecule of carbon dioxide (CO_2) and two molecules of water (H_2O).

Ways to Describe Chemical Reactions

Method	Example
Word Equation	methane* + oxygen ⟶ carbon dioxide + water
Chemical Equation	$CH_4 + 2\ O_2 \rightarrow CO_2 + 2\ H_2O$

*methane = natural gas

1. Which of the following events causes carbon dioxide and water to form?

(1) the reaction of methane and oxygen
(2) the mixture of water and carbon
(3) the evaporation of natural gas
(4) the reaction of carbon monoxide and oxygen
(5) a chemical change involving nitrogen

2. According to the table, which of the following is one effect of the burning of natural gas?

(1) air pollution
(2) formation of CO_2
(3) production of CH_4
(4) formation of O_2
(5) formation of carbon monoxide

Question 3 refers to the following passage.

In an ecosystem, the role of an individual species is called its niche. The niche of a species includes its interactions with other organisms. The habitat of a species refers to its physical location and normal surroundings. Each species in an ecosystem shares its habitat with many other species. One species is humans.

The North American robin, for example, has many habitats, including meadows, forests, parks, and pastures. Many other plants and animals share the robin's habitats. The robin's niche includes eating insects, worms, and fruit; nesting in trees; dispersing the seeds of fruits in its droppings; and fertilizing the soil around those seeds.

3. If humans destroy or move into more meadows, forests, and pastures, the habitat of the robin will probably shrink because of increased human activity. What effect is most likely the result of humans sharing more of the robin's habitat?

(1) a decreased population of robins
(2) an increase in forests and meadows
(3) an increase in soil fertilization
(4) increased food resources for robins
(5) greater variety of nesting sites

TIP

Watch for words that signal cause and effect: *because, as a result, due to, leads to, therefore, thus, so, the reason is,* etc.

Questions 4 and 5 refer to the following passage and illustration.

An estuary is a long bay at the mouth of a river. In an estuary, salty ocean water flows a certain distance upriver during high tide and retreats during low tide. The diagram shows the range of salinity that various estuary fish and shellfish can tolerate.

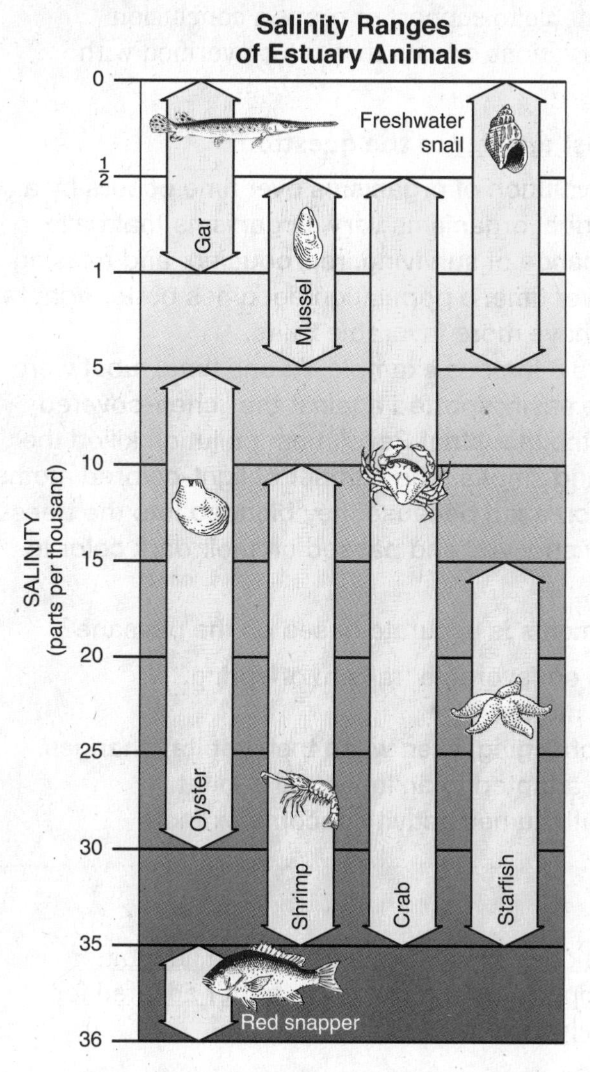

4. During periods of drought, the freshwater from the river decreases and ocean water tends to move farther into the estuary. Which event is a likely effect of drought?
   (1) Gar tend to migrate toward the sea.
   (2) Mussel population would increase.
   (3) Mussels and oysters share the same habitat.
   (4) Red snappers would move into the estuary.
   (5) Gar begin to feed on shrimp.

5. Global climate change is likely to cause unpredictable wide swings in salinity within estuaries as rainfall patterns and sea levels change. On which animal population are these shifts least likely to have an effect?
   (1) mussel
   (2) freshwater snail
   (3) shrimp
   (4) starfish
   (5) crab

Question 6 refers to the following passage.

Changes in matter that produce new substances are called chemical changes. Substances that enter into a chemical change are called reactants, and those that result from a chemical change are called products.

For example, when solid iron and sulfur are mixed together and heated, they undergo a chemical change to form solid iron sulfide. Iron and sulfur are the reactants, and iron sulfide is the product.

Another example of a chemical change occurs when an electric current is passed through water. Under certain conditions, the water changes chemically. It breaks down, and hydrogen gas and oxygen gas are produced. Water is the reactant, and hydrogen and oxygen are the products.

6. Which of the following causes the formation of hydrogen gas and oxygen gas?
   (1) changing iron through a chemical process
   (2) producing iron sulfide
   (3) mixing sulfur and water
   (4) passing an electric current through water
   (5) heating solid oxygen and hydrogen

Answers and explanations start on page 108.

Skill 10

Evaluation

Assess Adequacy and Accuracy of Facts

On the GED Science Test, you will **assess facts** in order to determine whether they are **adequate** and **accurate.** Check each choice against the information in the passage or the diagram to check for accuracy and adequacy.

- **Facts** are statements that are true and can be proven.
- An **adequate** fact is one that is suitable to support or prove a conclusion.
- An **accurate** fact is based on observations or data that can be verified with other facts.

Read the passage. Choose the one best answer to the question.

According to Charles Darwin, the evolution of organisms over time occurs by a process called natural selection. Individual organisms vary. Organisms that have more favorable traits have an added chance of surviving, reproducing, and passing on their favorable traits to offspring. Over time, a population becomes better adapted to its environment as more individuals have more favorable traits.

Peppered moths around London are a famous example. At one time, most were light-colored. The few dark moths were easily spotted against the lichen-covered tree trunks and eaten by birds. During the Industrial Revolution, pollution killed the lichen and coated the trees with soot and smoke. The number of light-colored moths declined. The number of dark moths increased because they blended into the trees and were difficult for birds to see. They survived and passed on their dark color to their offspring.

QUESTION: Which of the following statements is accurate based on the passage?

(1) All individuals of a species pass on favorable traits to offspring.
(2) Dark moths are superior to light moths.
(3) After species evolve, they stop changing, even when their habitat changes.
(4) Dark peppered moths were well adapted to an industrial habitat.
(5) Moths that come into contact with human activity become extinct.

EXPLANATIONS

STEP 1 To answer this question, ask yourself:

- What facts are given? survival of moth populations depended on adaptation
- What facts do the observations support? Dark moths were better adapted for industrial areas and were more likely to survive.

STEP 2 Evaluate all of the answer choices and choose the best answer.

(1) No. Not all individuals will reproduce, and those that do only have an added chance of passing favorable traits, not a guarantee.
(2) No. The passage implies only that certain moths are more likely to survive, not that one is superior.
(3) No. The passage says populations adapt over time.
(4) **Yes. This is accurate because dark moths were more likely to survive in an industrial environment because they matched their environment.**
(5) No. There are no facts supporting the extinction of a species of moths, only a decrease in the size of the population.

ANSWER: (4) Dark peppered moths were well adapted to an industrial habitat.

Practice the Skill

Try these examples. Choose the one best answer to each question. Then check your answers and read the explanations.

Questions 1 and 2 refer to the following passage and diagram.

Coal is a dark, carbon-rich rock mined as an energy resource. It is a nonrenewable resource formed from compressed plant remains over billions of years. Coal is highly combustible and releases high quantities of carbon dioxide when burned. The circle graph shows the uses of the coal mined in the United States.

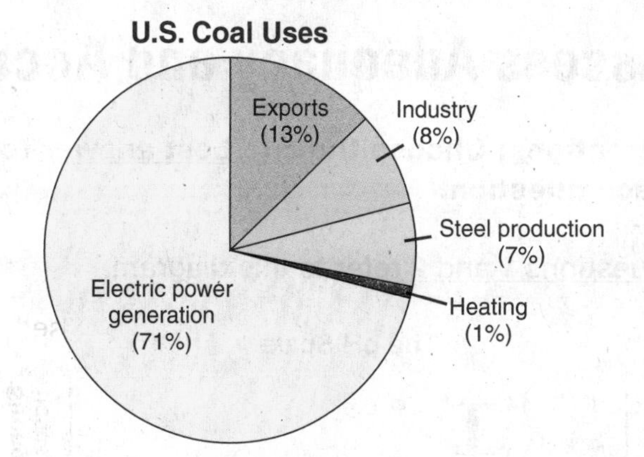

1. Which sentence is an accurate conclusion from the passage and graph?
   (1) Coal mined in the United States is sent to Canada, France, and England.
   (2) In this country, more coal is used for heating than for making steel.
   (3) Coal is a renewable resource.
   (4) Coal makes up the majority of U.S. exports.
   (5) Most of the coal mined in the United States is also used in the United States.

HINT **Make sure the data from the graph is accurately represented by your answer.**

2. Which fact adequately supports the conclusion that coal is an important source of energy in the United States?
   (1) More than 20% of coal is exported.
   (2) More than half the coal is used to generate electric energy.
   (3) More than 70% of coal is used for electricity in industrial applications.
   (4) Electricity is generated at power plants.
   (5) Two-thirds of known deposits of coal are found in the United States.

HINT **Which statement explains a fact that is not shown directly?**

Answers and Explanations

1. (5) Most of the coal mined in the United States is also used in the United States.
Option (5) is correct because the graph adequately shows the fact that only 13% of coal is exported.

There are no facts about destinations of exported coal (option 1). Option (2) is inaccurate because it has the facts reversed, as more coal is used for making steel than for heating. Option (3) contradicts the information in the passage, which states the coal is nonrenewable. While 13% of coal is exported, there is no information in the graph regarding how much of the country's total exports this amount is, so option (4) is incorrect.

2. (2) More than half the coal is used to generate electric energy.
Option (2) is correct because 71% is "more than half" of the graph, and this fact adequately supports the conclusion that electricity is important.

The amount of coal exported (option 1) is inaccurate—less than 20% is exported. Option (3) supports the conclusion, but it is not adequate by itself because the use of the electricity in industrial applications is not discussed. The fact that electricity is generated at power plants (option 4) is not represented in the passage or graph. The amount of coal found in the U.S. (option 5) is not represented in the passage or graph.

GED Practice

Assess Adequacy and Accuracy of Facts

Directions: Choose the one best answer to each question.

Questions 1 and 2 refer to this diagram.

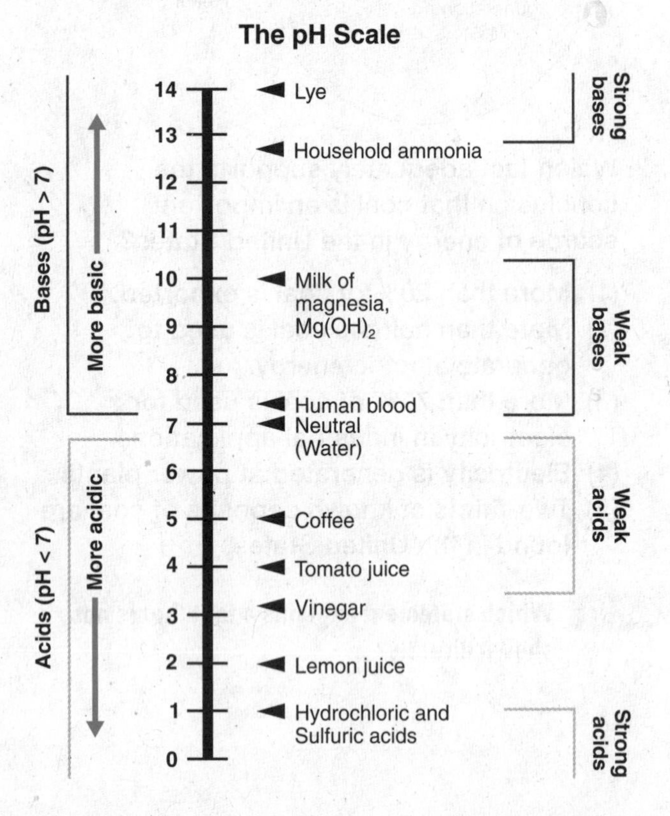

1. According to the diagram, which statement is accurate?
   (1) All beverages are strong acids.
   (2) Lemon juice is an acid and coffee is a base.
   (3) Coffee is more damaging to the stomach lining than lemon juice.
   (4) Coffee is less acidic than lemon juice.
   (5) Water is the only neutral substance.

2. Which sentence is an adequate fact for classifying human blood as a weak base?
   (1) Blood is more basic than coffee.
   (2) The pH of blood is between 7 and 10.
   (3) Blood is located near water on the chart.
   (4) Water is a major component of blood.
   (5) The pH of blood is greater than 7.

Questions 3 and 4 refer to the following passage.

To survive on Earth, we rely on substances called natural resources. There are two kinds of resources. A renewable resource can be replaced in nature after it is used. Renewable resources include air, sunlight, water, and plants.

Nonrenewable resources cannot be replaced. They are in limited supply because the processes that form them can take billions of years. Metals, minerals, coal, natural gas, and oil are all examples of nonrenewable resources.

3. Which of the following is an accurate statement that adequately supports the conclusion that trees are a renewable resource?
   (1) There are a large number of trees.
   (2) Trees only grow on Earth.
   (3) New trees can be planted to replace those that are cut down.
   (4) The processes that form renewable resources can take billions of years.
   (5) Trees require billions of years to grow.

4. Which of the following is an accurate statement that is adequately supported by the passage?
   (1) If we keep using Earth's nonrenewable resources, they will eventually be gone.
   (2) Metals and oil are renewable resources.
   (3) Nonrenewable resources are not distributed evenly on Earth's surface.
   (4) One of our nonrenewable resources comes from outside Earth.
   (5) All of our resources are produced only on Earth.

Question 5 refers to the following passage.

Humans have an effect on the evolution of many species. One example is the dandelion. Dandelions normally grow in undisturbed grassy fields with many other plants. Like other field plants, the dandelion must grow tall enough to get sunlight for photosynthesis. During their first year, field dandelions grow long leaves. They do not produce flowers and seeds until their second year.

Dandelions that grow in lawns have a different habitat. They are frequently mowed down or dug up by weeding gardeners. Lawn dandelions have adapted to this habitat in several ways. They bloom four months after they start to grow and produce three times as many seeds as field dandelions. Their root systems spread out underground. The stems of lawn dandelions are short and close to the ground. Their leaves are small and lie flat on the ground.

5. If closely mowed athletic fields were abandoned and left to grow wild, which of the following predictions would be accurate and adequately supported by the passage?
   (1) Very few plant species would grow on the fields.
   (2) Taller dandelions would begin to do well in the changed habitat.
   (3) The seeds of the dandelions would not be dispersed.
   (4) Tall dandelions would begin to bloom after three months.
   (5) Dandelions would not grow in the wild fields.

TIP

When asked to determine whether a statement is adequately supported by a passage, begin by ruling out any statements that contradict the passage. Then look for sentences or ideas in the passage that support the statement.

Question 6 refers to the following passage and diagram.

According to Darwin's theory of evolution, newer life-forms are modified forms of older life-forms. For example, different finches evolved from a recent common finch-like ancestor, as shown below.

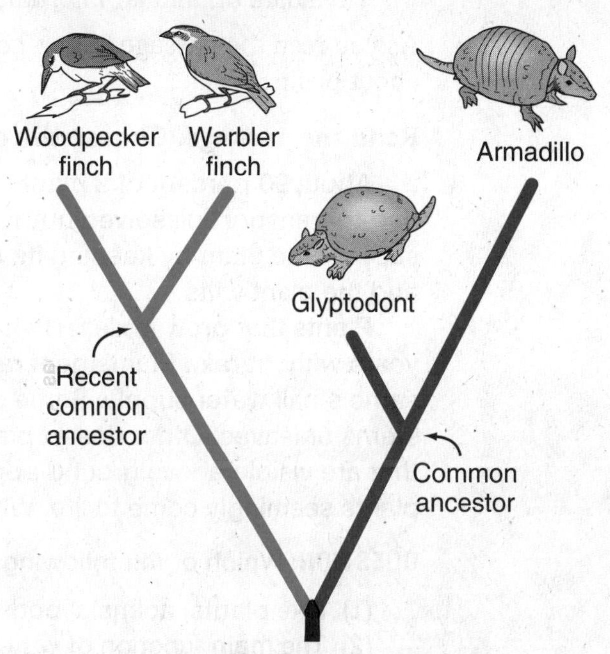

6. For which of these statements do the passage and diagram provide accurate support?
   (1) Armadillos evolved from finches.
   (2) Armadillos and glyptodonts share a recent common bird-like ancestor.
   (3) Armadillos and glyptodonts share a common armadillo-like ancestor.
   (4) Armadillos and finches have no common ancestors.
   (5) Finches evolved from glyptodonts.

Answers and explanations start on page 109.

Evaluation

Evaluate Information

On the GED Science Test, you may have to answer questions in which you must **evaluate information** for accuracy or logic.

- Evaluate familiar topics using your own **knowledge** and **experience.**
- Evaluate unfamiliar information using given facts and **logic.**

As you read the passage below, compare the information to what you already know about plants.

Read the passage. Choose the <u>one best answer</u> to the question.

About 90 percent of a plant consists of water. Water is needed for photosynthesis and to transport dissolved nutrients to each plant cell. In addition, water helps support the plant by keeping its cells rigid. Without enough water, the cells empty out and the plant wilts.

Plants that grow in deserts have developed ways of surviving for months or even years without rain. Plants generally grow far apart so they do not compete for the same small water supply. Some desert plants, such as cacti, store water in thick stems or leaves. Other desert plants store water in their roots. The parts of the plant that are visible above ground appear dried out and dead. When rain finally falls, the plants seemingly come to life. Within weeks, they grow rapidly, flower, and reproduce.

QUESTION: Which of the following statements is supported by the passage?

(1) Like plants, animals' body weight consists mostly of water.
(2) The main function of water in a plant is to provide rigidity.
(3) Cacti store water in their roots, stems, and leaves.
(4) One purpose of a cactus's needles is to protect it from predators.
(5) Many desert plants have a very brief growing season.

EXPLANATIONS

STEP 1 To answer this question, ask yourself:

- What information does the passage provide? <u>how plants adapt to minimal water</u>
- How can I evaluate the statements? <u>compare the information to facts in the passage and choose the one that logically matches</u>

STEP 2 Evaluate all of the answer choices and choose the <u>best</u> answer.

(1) No. The passage does not have information about animals.
(2) No. This is one function of water, but not the main function.
(3) No. According to the passage, cacti only store water in stems and leaves, not the roots.
(4) No. The passage does not mention cactus needles.
(5) **Yes. The passage states that the plants have long periods of no growth followed by a rapid growth cycle following rain.**

ANSWER: (5) Many desert plants have a very brief growing season.

Practice the Skill

Try these examples. Choose the one best answer to each question. Then check your answers and read the explanations.

Questions 1 and 2 refer to the following passage and graph.

Earth's seasons are caused by the tilt of the planet's axis. In the summer, the Northern Hemisphere tilts toward the sun; the sun's rays hit Earth at a more direct angle, and the temperatures rise. In the winter, the Northern Hemisphere tilts away from the sun; the sun's rays hit Earth at an indirect angle, and the temperatures are cooler.

The table shows the high and low temperatures for several cities in the Southern United States.

Temperatures for Selected Southern Cities				
	Today's Forecast		Yesterday	
City	High	Low	High	Low
Atlanta	60	42	68	51
Charleston	66	46	80	52
Charlotte	58	41	74	52
Jackson	50	36	69	57
Louisville	42	29	65	50
Memphis	49	30	67	51
Nashville	47	35	62	51
Orlando	79	62	83	64
New Orleans	56	45	76	59

1. Based on your knowledge and the information provided, which of the statements below is a reasonable conclusion?

(1) All of the cities will be warmer tomorrow than they are today.
(2) Rain is likely across most of the south today.
(3) These temperatures were recorded sometime between late fall and early spring.
(4) Ice may hinder navigation along the Mississippi River near Memphis.
(5) No city will have a temperature greater than the forecasted high.

HINT **What do you know about weather forecasts and the geography of the U.S.?**

2. According to the table, the widest temperature range over the two-day period will occur in which city?

(1) Charleston
(2) Charlotte
(3) Memphis
(4) Orlando
(5) New Orleans

HINT **How do you determine the temperature range using the data in the table?**

Answers and Explanations

1. (3) These temperatures were recorded sometime between late fall and early spring.
Option (3) is correct; these relatively cool temperatures are only likely in the Southern United States during winter.

Option (1) is incorrect; you cannot predict a trend from two days' data. Option (2) is incorrect; there is no data regarding rainfall. Option (4) is incorrect; the warm temperatures yesterday make this answer unlikely. Option (5) is incorrect; temperature forecasts are not always accurate.

2. (3) Memphis
Option (3) is correct because the temperature range is the difference between the highest and lowest temperatures during a period of time. The range for Memphis was 67° – 30° = 37°.

The other cities all had smaller ranges during the period: Charleston, 34° (option 1); Charlotte, 33°(option 2); Orlando, 21° (option 3); New Orleans, 31° (option 4).

Evaluate Information

Directions: Choose the one best answer to each question.

Questions 1 and 2 refer to this passage and diagram.

In flowering plants, the male stamens produce pollen. The male stamens contain the male sex cells. When pollen is transferred from the stamen to the female stigma, it travels down the pollen tube to the ovary. Fertilization takes place when the male sex cells join with the female sex cells, and seeds start to grow.

Although some plants can pollinate themselves, most do not. Often self-pollination cannot occur because the stamens and stigmas mature at different times. This timing helps to ensure the genetic diversity of the species.

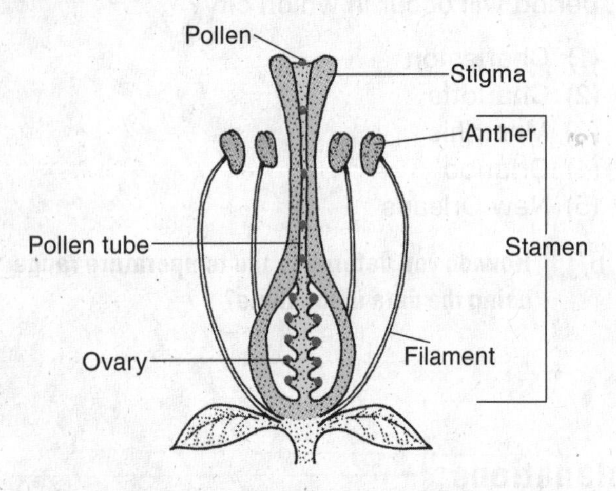

1. Which statement is accurate based on information in the passage and the diagram?
   (1) Flowering plants reproduce by transferring pollen from the female stigma to the male stamens.
   (2) The anther contains the female sex cells of flowering plants.
   (3) Pollen rises up through the pollen tube and is transferred to the stigma.
   (4) Male sex cells of a flowering plant are produced in the ovary.
   (5) Flowering plants produce seeds after fertilization takes place.

2. Based on the information in the passage, the diagram, and what you already know, which of these statements is a fact?
   (1) Flowers play an important role in the sexual reproduction of plants.
   (2) Pollen from any flower can fertilize the female cells of any other flower.
   (3) The ovary of a flower is part of the male reproductive organ of a plant.
   (4) The pollen tube is part of the stamen.
   (5) All plants require insects for pollination to occur.

Question 3 refers to the following passage.

Roots anchor plants in the soil and provide a stable foundation for growth by branching in all directions to hold the plant firmly. They also absorb water and minerals from the soil. Water and minerals pass from the soil into the tiny hairs on the younger parts of the roots. Then water and minerals move into the center part of the roots, where they are carried up to stems and leaves. Some plants, such as trees, dandelions, and carrots, have a thick main root called a taproot, which is used to store sugar or starch. The plant uses this stored food to survive during the winter and to produce new growth in the spring. Other plants, such as grasses, have a network of many thin, fibrous roots.

3. Evaluate the statements below and choose the one that is best supported by the passage.
   (1) Plants absorb rainwater through openings in their leaves.
   (2) Plant roots grow with such force that they can crack rock.
   (3) When a potted plant is placed on its side, new roots grow downward.
   (4) Two basic types of root systems are taproots and fibrous roots.
   (5) Excess water not needed by the plant passes from the roots into the soil.

Questions 4 through 6 refer to the following passage and diagram.

A wave transfers energy from one place to another without transferring matter. When a mechanical wave travels through a medium, the particles of the medium move up and down or back and forth, but they do not move in the same direction as the wave. For example, think about "the wave" at a sports stadium. A wave moves around the circumference of the stadium; however, each person stays in the same seat. The diagram below shows several characteristics used to describe waves traveling through a medium: crest, trough, amplitude, and wavelength. The number of wavelengths that pass a given point in one second is called the frequency.

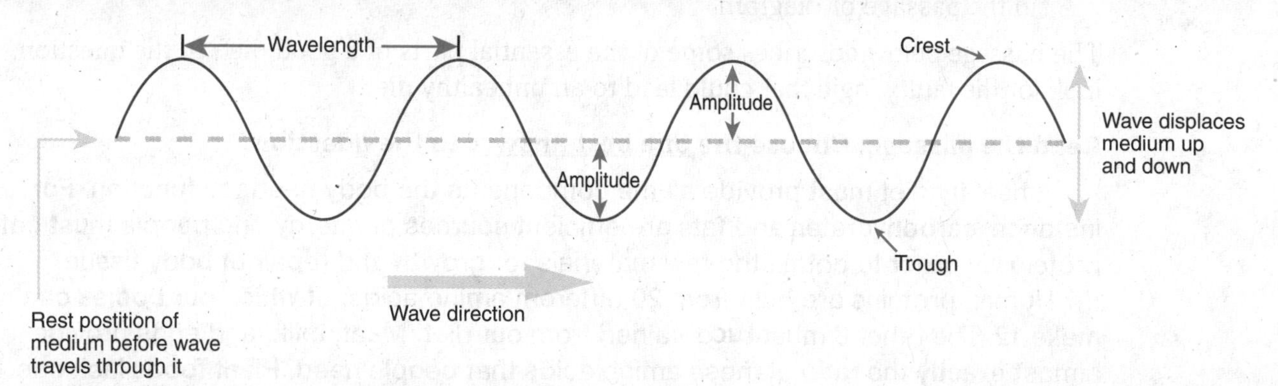

4. Evaluate the information in the passage and the diagram. Which of the following statements can be concluded?
   (1) The wavelength and the frequency of a wave are two terms for the same property.
   (2) The greater the energy of a wave is, the smaller the wave's amplitude is.
   (3) The amount of displacement of the medium depends on amplitude, not on wavelength.
   (4) A wave's amplitude is always longer than its wavelength.
   (5) Light waves travel faster than sound waves.

5. According to the information in the diagram, what is the distance that the medium moves, measured from the rest position to the highest or lowest point on the wave?
   (1) trough
   (2) crest
   (3) wavelength
   (4) amplitude
   (5) frequency

6. Which of these statements does the diagram support?
   (1) The distance from one trough to the next is one wavelength.
   (2) Waves travel from left to right.
   (3) A change in wavelength causes an identical change in amplitude.
   (4) Particles of the medium travel along the wave exactly one-half wavelength and then return to their rest position.
   (5) The ratio of crests and troughs in a wave is 4:3.

TIP

When you evaluate information, look for all the facts that are presented in the passage, graph, or diagram. Also, take into account facts that you already know about the topic and figure out how they relate to the new information.

Answers and explanations start on page 109.

Evaluation

Recognize Faulty Logic

On the GED Science Test, you may have to answer questions in which you must **recognize faulty logic** that leads to a conclusion that does not make sense.

- To find faulty logic, look for details that **do not support** the conclusion.
- Look for **cause-and-effect** relationships that do not really exist or are not justified in the passage or diagram.

The passage below describes some of the essential parts of a good diet. In the question, look for the faulty logic that could lead to an unhealthy diet.

Read the passage. Choose the one best answer to the question.

A healthy diet must provide all the components the body needs to function. For instance, carbohydrates and fats are efficient sources of energy and people must eat proteins in order to obtain the raw materials for growth and repair of body tissue.

Human proteins are built from 20 different amino acids, of which our bodies can make 12. The other 8 must be obtained from our diet. Meat, milk, and eggs provide almost exactly the ratio of these amino acids that people need. Plant foods lack one or more of the essential amino acids, so vegetarian diets must be planned to provide all of the amino acids. Grains tend to lack particular amino acids that are common in legumes, such as beans. Leafy green vegetables have all but one of the essential amino acids.

QUESTION: A woman decided that carbohydrates could provide all the energy she needed, so she stopped eating meat, vegetables, and sweet desserts. Her diet included a variety of breads, pastas, and whole grains. After three weeks, she stopped the diet because she didn't feel well. What was faulty about her logic in designing the high-energy diet?

(1) The diet contained lots of grain products, so it provided too much energy.
(2) Desserts were not permitted, so the woman did not get enough amino acids.
(3) The meal plan did not vary, so the diet became boring and hard to follow.
(4) The meal plan was not flexible, so the diet was hard to follow when eating out.
(5) The diet contained mostly grain products, so it lacked essential amino acids.

EXPLANATIONS

STEP 1 To answer this question, ask yourself:

- What information does the passage provide? elements of a complete diet
- What does a healthy diet require? a balance of the different types of amino acids

STEP 2 Evaluate all of the answer choices and choose the best answer.

(1) No. There is no indication that too much energy will make you feel bad.
(2) No. Desserts are not likely to be good sources of amino acids.
(3) No. Even if the woman thought the diet was boring, this does not explain why she did not feel well.
(4) No. This is a drawback of many diets, but it is not related to health.
(5) **Yes. A diet that relies completely on grains lacks amino acids that are needed to build and repair tissues.**

ANSWER: (5) The diet contained mostly grain products, so it lacked essential amino acids.

Practice the Skill

Try these examples. Choose the <u>one best answer</u> to each question. Then check your answers and read the explanations.

<u>Questions 1 and 2</u> refer to the following passage and diagram.

In plants, growth occurs only in special tissues called meristems. Apical meristems are located at the tips of stems, branches, and roots, where the plant grows longer.

The main stem grows from the terminal bud at its end, which contains the apical meristem. As the tip grows, it leaves behind some meristem, which grows into lateral buds. Lateral buds may grow into branches, flowers, or leaves. Sometimes lateral buds do not grow until the terminal bud on the main stem has stopped growing.

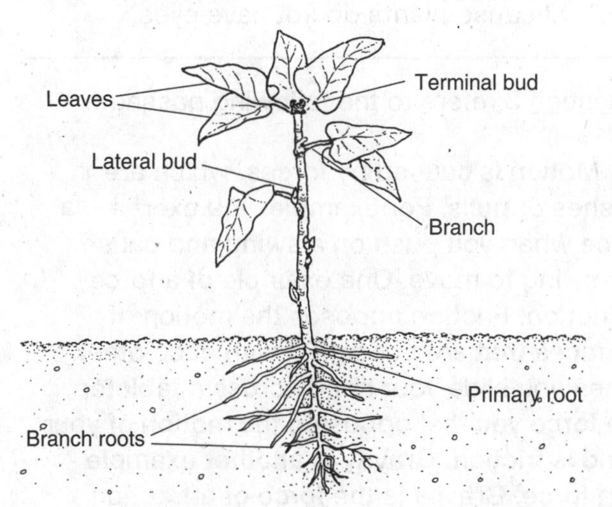

1. Which of the following statements uses faulty logic?

   (1) Apical meristems allow plants to grow longer.
   (2) Lateral buds can grow into several different things.
   (3) Lateral buds control a plant's height.
   (4) Lateral buds can depend on terminal bud activity.
   (5) Meristems are necessary for a plant's survival.

HINT **Which choice is not supported by the information in the passage?**

2. Which of the following statements is logically supported by the information given in the passage and diagram?

   (1) Bushy plants have many terminal buds.
   (2) The terminal bud controls the height of the plant's main stem.
   (3) Without a terminal bud, the plant's roots do not grow.
   (4) Lateral buds grow before terminal buds.
   (5) A tall plant always has many branches.

HINT **Where is the terminal bud located on the plant?**

Answers and Explanations

1. (3) Lateral buds control a plant's height.
Option (3) is correct. According to the passage, terminal buds control a plant's height, and lateral buds grow branches, flowers, or leaves.

Options (1) and (2) are logical and accurate based on the information in the passage. Option (4) is logical; the passage says that some lateral buds do not start growing until the terminal bud stops growing, so it is logical to say that lateral buds can depend on terminal bud activity. Option (5) is logical; growth in plants only occurs in meristems, so it is logical that a plant needs meristems to survive.

2. (2) The terminal bud controls the height of the plant's main stem.
Option (2) is correct because the terminal bud is the growth point of the main stem.

Bushy plants grow in all directions, so they have many lateral buds (option 1), not terminal buds. Terminal buds are not related to root growth (option 3). Option (4) is incorrect; the passage states that lateral buds often do not grow until the terminal bud stops growing. If the terminal bud is the only growing bud, a tall plant may actually have only a few branches (option 5).

GED Practice

Recognize Faulty Logic

Directions: Choose the one best answer to each question.

Questions 1 and 2 refer to the following passage and graph.

Part of the study of genetics involves being able to predict the appearance (phenotype) and genetic makeup (genotype) of the offspring of organisms. Geneticists have identified three different genes that play a role in eye color: the green/blue eye color gene, the brown eye color gene, and the brown/blue eye color gene.

Each gene has several different forms, or alleles. When a person has two different alleles for a particular gene, one of the alleles wins out over the other. This allele is labeled as the dominant allele. The graph below shows the distribution of eye colors among a population.

Eye Color of 100 People Surveyed

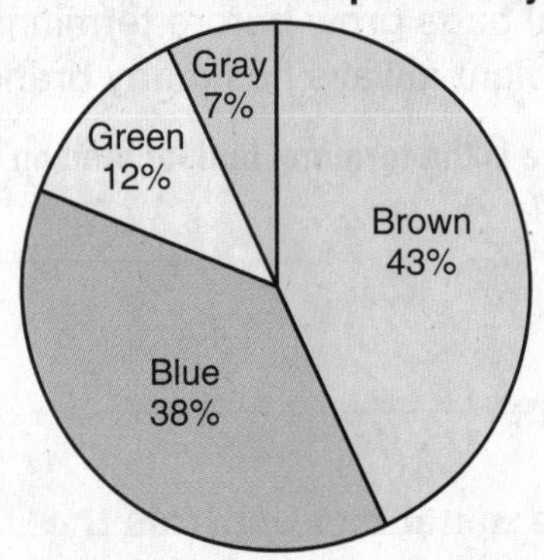

1. Which of these statements does not demonstrate faulty logic?
   (1) Eye color is determined by a single gene because both eyes are the same color.
   (2) Blue is a dominant eye color because there are two genes responsible for blue eyes.
   (3) Brown alleles are most likely dominant because most people have brown eyes.
   (4) There are only four possible eye colors, as shown by the graph.
   (5) Blue eyes cannot be dominant because they are the lightest color.

2. Based on the information provided, which of the following is a logical conclusion?
   (1) The phenotype of a trait is determined by the dominant gene.
   (2) An organism can only have one allele for each gene because it shows only one trait.
   (3) Brown or blue eyes are better than green eyes because they are more common.
   (4) Eventually all people will have brown eyes because it is the dominant gene.
   (5) Only animals and humans have alleles because plants do not have eyes.

Question 3 refers to the following passage.

Motion is caused by forces, which are pushes or pulls. For example, you exert a force when you push on a swing and cause the swing to move. One example of a force is friction. Friction opposes the motion of surfaces that are in contact with each other. When you slide your hand across a tabletop, the force you feel opposing the motion of your hand is friction. Gravity is another example of a force. Gravity is the force of attraction possessed by all masses. Every mass attracts every other mass with this force, which is a function of mass and distance. Earth's gravity acts on a ball you throw in the air. Gravity causes it to fall to the ground.

3. Which statement is a logical conclusion based on the passage?
   (1) Moving objects are not affected by friction if they do not stop.
   (2) Friction does not affect a ball because it rolls rather than slides.
   (3) If two objects do not move toward one another, there is no gravitational force between them.
   (4) The motion of an object is affected by all the forces that act on it.
   (5) Gravity is a stronger force than friction because it causes things to fall.

Questions 4 and 5 refer to the following passage and diagram.

Lightning occurs when electrically charged particles travel between areas of opposite charge. Lightning strikes the ground, or an object on the ground, when a negatively charged area forms at the base of a cloud and objects on the ground below become positively charged. Electrons in the cloud are attracted to the ground and a stream of them begins moving downward.

The flow of electrons releases a large amount of energy, some of which is converted to light and sound. The charges are able to build because the air generally acts as an electrical insulator. Lightning does not occur until the difference in electrical potential is very high.

4. Which of the following statements is a logical conclusion based on the diagram and passage?
   (1) Lightning only occurs when it is raining.
   (2) Most of the energy of lightning occurs as light and sound.
   (3) Lightning does not occur between two objects with the same electrical charge.
   (4) If the clouds are moving, there cannot be any lightning because charges cannot build.
   (5) Lightning is not dangerous because it is only an electric current.

5. Based on the passage and the diagram, which of the following statements demonstrates faulty logic?
   (1) Lightning can occur between two regions inside a cloud.
   (2) Lightning travels from the areas marked with a negative sign to the areas marked with a positive sign.
   (3) A tree is not a safe place to be during a thunderstorm because lightning is likely to strike taller objects.
   (4) The energy of lightning is caused by a flow of charged particles.
   (5) Lightning cannot strike the tree because there is a positive charge on the tree.

Question 6 refers to the following passage.

The United States was the first country to use nuclear-powered submarines. Previous submarines used batteries and diesel fuel; these subs had to frequently resurface to recharge the batteries and get oxygen. Nuclear fission reactors produce controlled nuclear reactions; these reactions release large amounts of energy without using any oxygen. These reactors provide subs with energy for long periods of time without needing to recharge or refuel.

6. Which of the following statements logically supports the conclusion that a nuclear sub is capable of exploring below the vast ice of the Arctic Ocean?
   (1) Nuclear reactors provide energy.
   (2) Fission reactors produce heat.
   (3) Nuclear-powered subs can remain submerged for a much longer time.
   (4) The United States was the first country to use a nuclear submarine.
   (5) Diesel subs sometimes rely on batteries.

TIP

When looking for faulty logic, ask "Is the outcome supported by the details?"

Answers and explanations start on page 110.

Graphic Skills

Tables and Charts

On the GED Science Test, you will have to answer questions in which you interpret data that is presented in **tables and charts.**

- A table or chart usually organizes information in rows and columns.
- To find information, look under the **column** heading, and scan down the column until you locate the **row** that identifies the information you need.

Examine the graph. Choose the <u>one best answer</u> to the question.

Comparing Acids and Bases

	Acids	Bases
Ion Formation in Water	give hydrogen ions (H^+) to water molecules to form hydronium ions (H_3O^+)	accept hydrogen ions (H^+) from water molecules to form hydroxide ions (OH^-)
Properties	corrode metals, turn litmus paper red, taste sour	corrode hair and wool, turn litmus paper blue, taste bitter
Examples	hydrochloric acid, sulfuric acid	ammonia, sodium carbonate

QUESTION: Which of the following statements correctly identifies a similarity between acids and bases?

(1) Acids and bases have a similar taste.
(2) Acids and bases corrode metals, hair, and wool.
(3) Acids and bases both form ions in water.
(4) Acids and bases affect litmus paper in the same way.
(5) Acids and bases are both used in swimming pool chemicals.

EXPLANATIONS

STEP 1 To answer this question, ask yourself:

- What information is in the columns of the chart? <u>properties of acids and bases</u>
- Where should you look for similarities? <u>in the same row</u>

STEP 2 Evaluate all of the answer choices and choose the <u>best</u> answer.

(1) No. The properties row indicates that the tastes are different.
(2) No. The chart indicates acids corrode metals and bases corrode hair and wool.
(3) Yes. The ion-formation row indicates that both form ions in water.
(4) No. The chart indicates that acids and bases affect litmus paper differently.
(5) No. The table does not have any information about the use of acids and bases in swimming pools, so this statement is not correct.

ANSWER: (3) Acids and bases both form ions in water.

Practice the Skill

Try these examples. Choose the one best answer to each question. Then check your answers and read the explanations.

Questions 1 and 2 refer to the following passage and table.

The four major blood groups are A, B, AB, and O. Blood groups are defined by the presence of antigens and antibodies in blood plasma. Red blood cells have unique antigens that give blood cells their identity. Plasma contains antibodies produced by the body to destroy foreign antigens. Blood type determines which antibodies exist. For example, a person with type A blood carries the A antigen and B antibodies. If a blood transfusion is made with blood containing an antigen for which the recipient produces antibodies, a potentially lethal rejection occurs.

Blood Types

Blood Type	Antigen		Antibody	
	A	B	anti-A	anti-B
A	+	–	–	+
B	–	+	+	–
O	–	–	+	+
AB	+	+	–	–

(+) indicates protein's presence
(–) indicates protein's absence

1. Based on the table, which of the following statements is true?
   (1) Type AB blood has A and B antibodies.
   (2) People with the B antigen do not have B antibodies.
   (3) People with the A antigen have both A and B antibodies.
   (4) People who have neither the A nor the B antigen have neither A nor B antibodies.
   (5) People with the A and B antigens have A and B antibodies.

HINT **Compare data in the same rows of the antigen and antibody columns.**

2. According to the passage and the table, if someone with an unknown blood type requires a transfusion, what blood type would be the safest to give?
   (1) type A
   (2) type B
   (3) type O
   (4) type AB
   (5) any blood type

HINT **Remember that an antigen cannot mix with an antibody of the same classification.**

Answers and Explanations

1. (2) People with the B antigen do not have B antibodies.
Option (2) is correct because the type B row shows the presence of the B antigen and absence of the B antibody.

People with type AB blood (option 1) have two negative signs in the antibody columns, which means the antibodies are not present. Option (3) is incorrect; the table shows that a person with the A antigen never has the A antibody. People with neither antigen have both antibodies (option 4), while people with both antigens do not have either antibody (option 5).

2. (3) type O
Option (3) is correct because the passage says that matching wrong antigens can be lethal. The table shows that type O blood does not contain either antigen but does contain both antibodies, so it can be mixed safely with any combination of antibodies.

Type A blood (option 1) is not compatible with a type B or O recipient because they contain incompatible antibodies. Type B blood (option 2) is not compatible with type A or O because they contain incompatible antibodies. Type AB blood (option 4) is not compatible with any recipient other than type AB. Option (5) is incorrect because the antibodies of types A, B, and AB would not be safe.

Tables and Charts

Directions: Choose the one best answer to each question.

Questions 1 and 2 refer to the following passage and chart.

Taxonomy is the branch of science that develops the rules for naming, describing, and classifying organisms into different categories. The biological sciences use a classification system invented by Carolus Linnaeus. The Linnaean classification system organizes all living things according to a prescribed hierarchy.

Biologists classify organisms into the different categories by assessing the similarities and differences of each sample—the more similarities that exist, the closer the biological relationship of the organisms.

Classification of Living Things in the Linnaean System

Kingdom	Structure	Nutrition	Types of Organisms
Protista	Large, single-celled organism	Absorb, ingest, or photo-synthesize food	Protozoans and algae
Fungi	Multicellular organism	Absorb food	Fungi, molds, mushrooms, yeasts, mildews, smuts
Plantae	Multicellular organism	Photo-synthesize food	Mosses, ferns, non-flowering and flowering plants
Animalia	Multicellular organism	Ingest food	Sponges, worms, insects, fish, birds, amphibians, reptiles, mammals

1. According to the chart, which of the following organisms share the most similarities with mammals?
   (1) bacteria
   (2) protozoans
   (3) plantae
   (4) amphibians
   (5) fungi

2. How do ferns, mosses, and non-flowering plants obtain their nutrition?
   (1) by absorbing food from the environment
   (2) by ingesting nutrients from the ground
   (3) by photosynthesizing food
   (4) by ingesting airborne food particles
   (5) by absorbing, ingesting, and photo-synthesizing food

Question 3 refers to this chart.

Typical Function of Some Endocrine Glands

Gland	Activity Regulated
Pituitary	growth; regulates other glands
Thyroid	metabolic rate (body weight)
Thymus	lymphatic system (immunity)
Parathyroid	calcium metabolism (nervous system)
Pancreas	insulin production (sugar metabolism)
Adrenal cortex	salt & carbohydrate metabolism

3. Diabetes is caused by the insufficient production of insulin. Based on the table, a malfunction in which of the body's endocrine glands might lead to diabetes?
   (1) thymus
   (2) pancreas
   (3) parathyroid
   (4) thyroid gland
   (5) adrenal cortex

Questions 4 through 6 refer to the following table.

The table below shows earthquake magnitudes, measured by seismographs using the Richter Scale, and frequencies for all of Earth, as well as their damaging effects.

Magnitude on Richter Scale	Number per Year	Effects in Populated Areas
Less than 3.4	800,000	recorded only by seismographs
3.5 to 4.2	30,000	felt by some people
4.3 to 4.8	4,800	felt by many people
4.9 to 5.4	1,400	felt by everyone in the area
5.5 to 6.1	500	slight building damage
6.2 to 6.9	100	much building damage
7.0 to 7.3	15	serious damage: bridges twisted, walls fractured
7.4 to 7.9	4	great damage: buildings collapse
More than 8.0	1 per 5 to 10 years	total damage

4. Which of the following sentences is most likely a description of an earthquake with magnitude 4.9 to 5.4?
   (1) Everyone in Santa Cruz feels the ground shake, but no major damage occurs.
   (2) Some people in Fresno feel the ground shake.
   (3) Many people in San Francisco feel the ground shake, but no damage occurs.
   (4) A bridge across the bay becomes twisted.
   (5) Many buildings in Los Angeles collapse.

5. Which of the following is most similar to the effect of an earthquake measuring 7.4 to 7.9 on the Richter Scale?
   (1) A family feels the ground shake as a truck drives by.
   (2) A wrecking ball flattens a row of old houses.
   (3) Only two out of ten people feel the ground shake as a train goes by.
   (4) A falling cabinet causes a wall to fracture.
   (5) During a storm, all the windows in a house get broken.

6. Which of the following is an example of a use for the Richter Scale?
   (1) predicting the minimum number of earthquakes Bombay will get each century
   (2) counting the number of buildings damaged in an earthquake
   (3) comparing the strength of earthquakes in Tokyo and Calcutta
   (4) ranking the effectiveness of seismographs throughout California
   (5) comparing the speed of earthquake waves

TIP

Always identify what specific information is needed to answer the question—tables and charts often contain more information than is needed. Carefully read and understand the topic of the chart and the headings of the columns and rows. Identify where the necessary data is located by following the appropriate column and row to where they intersect.

Answers and explanations start on page 110.

Graphic Skills

Bar Graphs

On the GED Science Test, you may have to answer questions in which you interpret data that is presented in a **bar graph.**

- A **bar graph** is used to compare numerical data.
- Relative values are shown by the height of a **vertical bar** or the length of a **horizontal bar.** A tall or long bar indicates a large value. A short bar indicates a small value.

The bar graph below presents information about seed storage temperatures.

Examine the graph. Choose the one best answer to the question.

Green dragon plants grow all over North America. The bar graph to the right compares germination rates of green dragon seeds that were collected in Canada and Louisiana and were stored at two different temperatures.

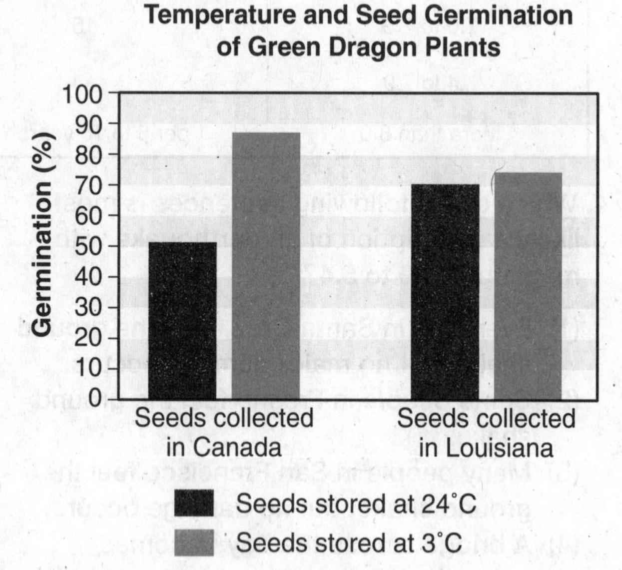

QUESTION: Which seeds have the highest germination rate?

(1) seeds from Canada stored at 24°C
(2) seeds from Canada stored at 3°C
(3) seeds from Louisiana stored at 24°C
(4) seeds from Louisiana stored at 3°C
(5) All seeds germinated at the same rate.

EXPLANATIONS

STEP 1 To answer this question, ask yourself:

- What does the height of the bars represent? the percent of seeds that germinate
- Which bar provides the answer to the question? the tallest bar

STEP 2 Evaluate all of the answer choices and choose the best answer.

(1) No. The bar for Canadian seeds stored at 24°C is shorter than the other bars, which means it has the lowest germination rate.
(2) **Yes. This group of seeds has a rate of about 85% germination, which is the highest rate, and it is represented by the tallest bar on the graph.**
(3) No. The bar for this group indicates about 70% germination, less than that of Canadian seeds stored at 3°C.
(4) No. The bar representing this group is taller than the bar beside it but shorter than the Canadian seeds stored at the same temperature.
(5) No. The bars have different heights, indicating different germination rates.

ANSWER: (2) seeds from Canada stored at 3°C

Practice the Skill

Try these examples. Choose the one best answer to each question. Then check your answers and read the explanations.

Questions 1 and 2 refer to this bar graph and passage.

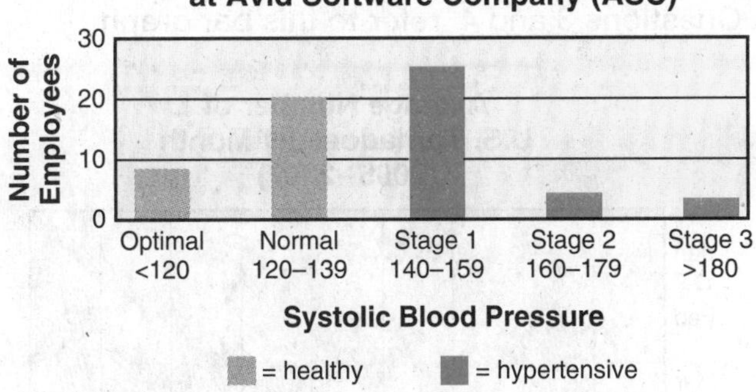

Blood pressure is measured in millimeters of mercury (mm Hg). Research indicates that, for the best health, systolic blood pressure (the first and higher number in a blood pressure reading) should be less than 120.

1. According to the bar graph, about how many employees at ASC exhibit blood pressure in the hypertensive range?

   (1) 3
   (2) 4
   (3) 8
   (4) 32
   (5) 52

HINT **Which bars on the graph indicate hypertension?**

2. According to recent estimates, one in four U.S. adults has high blood pressure. Which of the following conclusions is supported by the bar graph?

   (1) Jobs at ASC are stressful, so many employees have high blood pressure.
   (2) A larger percentage of ASC employees have high blood pressure than the population average.
   (3) Fewer employees than expected at ASC have high blood pressure based on the average U.S. population.
   (4) All ASC employees exhibit symptoms of high blood pressure.
   (5) Half of the employees at ASC have stage 1 blood pressure.

HINT **Read answers carefully to determine what comparisons are being made.**

Answers and Explanations

1. (4) 32

Option (4) is correct because according to the graph, stages 1, 2, and 3 show hypertension. The graph shows 25 stage 1 results, four stage 2 results, and three stage 3 results, and 25 + 4 + 3 = 32.

The values of 3 employees (option 1) and 4 employees (option 2) refer to one of the three stages that are hypertensive, but not the total. The value of 8 employees (option 3) refers to the number of employees with optimal pressures. Option 5 includes the 20 employees in the normal range.

2. (2) A larger percentage of ASC employees have high blood pressure than the population average.

Option (2) is correct because more than 50% of employees have high blood pressure compared to 25% of the U.S. population.

There is no information about the cause (option 1) of high blood pressure at ASC. Option (3) is contradicted by the data. Almost half of the employees have normal blood pressure, so option (4) is incorrect. Stage 1 (option 5) is the highest bar but is less than 50% (or half) of the total.

GED Practice

Bar Graphs

Directions: Choose the one best answer to each question.

Questions 1 and 2 refer to this bar graph.

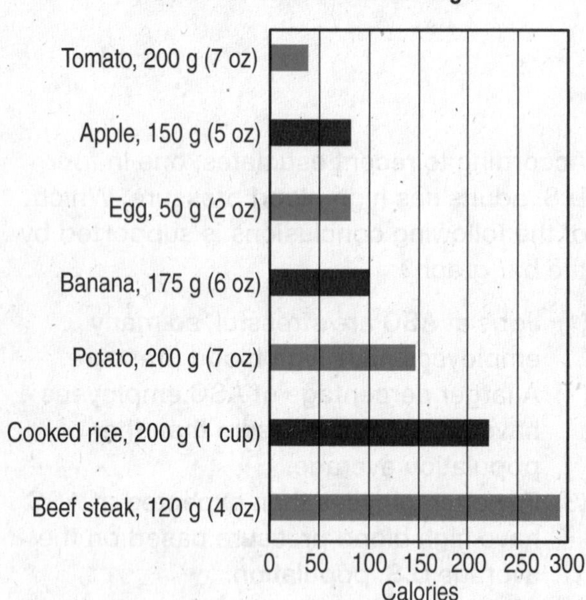

1. The average 30-year-old woman burns about 75 calories during a 1-mile walk. Which food would a woman in this category be able to eat and then burn off the calories during her walk?

   (1) apple
   (2) banana
   (3) potato
   (4) cooked rice
   (5) beef steak

2. According to the graph, which of the following foods has the fewest calories per serving?

   (1) beef steak
   (2) tomato
   (3) cooked rice
   (4) potato
   (5) banana

Questions 3 and 4 refer to this bar graph.

Average Number of U.S. Tornadoes by Month (2005–2007)

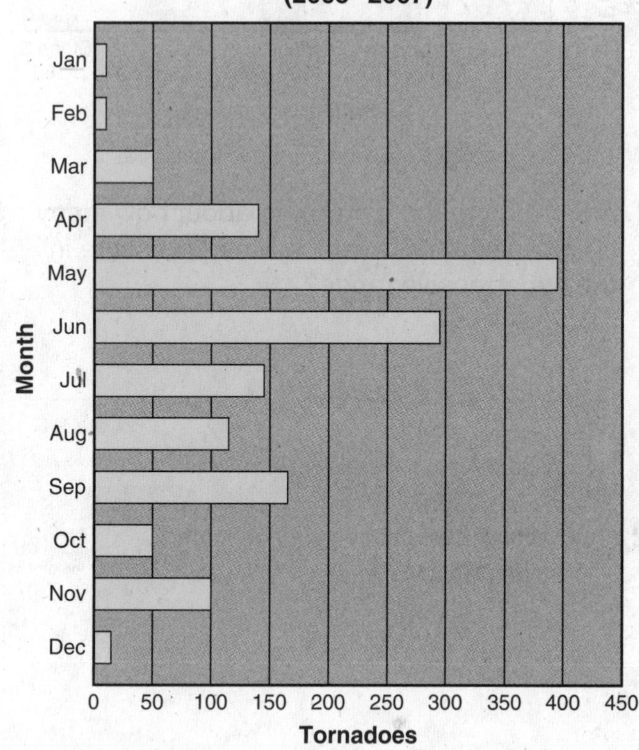

3. During which season would scientists be able to gather the most information about the formation of tornadoes?

   (1) winter (Jan–Mar)
   (2) spring (Apr–Jun)
   (3) summer (Jul–Sep)
   (4) autumn (Oct–Dec)
   (5) none

4. Which is the best estimate of the total number of tornadoes that occur during the months of May, June, and July?

   (1) 280
   (2) 400
   (3) 700
   (4) 850
   (5) 1,500

Questions 5 through 8 refer to this bar graph.

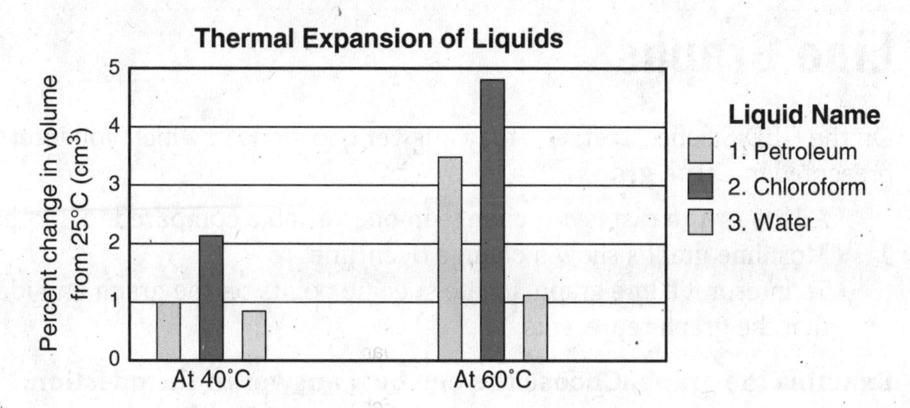

5. Which statement accurately describes the information in the graph?
   (1) The average change in volume for all liquids is the same for both temperatures.
   (2) Volume increased in all three liquids from 40°C to 60°C.
   (3) Water was the only liquid to experience a decrease with increased heat.
   (4) The change in volume from 40°C to 60°C was negligible.
   (5) Chloroform is more expensive than petroleum.

6. Which of the following statements is true based on the graph?
   (1) Expansion by heat is dependent on the kind of liquid.
   (2) Gases expand more when heated than liquids at the same temperature.
   (3) The volume of a liquid is independent of temperature.
   (4) Petroleum and chloroform have the same percent of volume increase.
   (5) Liquids will explode at 80°C.

7. A 100 cm³ sample of an unknown liquid at 25°C is heated to 55°C. What will most likely happen to the volume of the liquid?
   (1) no change
   (2) increase by 3%
   (3) increase by 20%
   (4) increase by an unknown amount
   (5) decrease by an unknown amount

8. Based on the information in the graph, which liquid is least affected by changes in teperature?
   (1) all
   (2) petroleum
   (3) chloroform
   (4) water
   (5) none

TIP

When you read a graph, carefully read the title and the labels. These provide important information about what data is or is not included in the graph.

Answers and explanations start on page 111.

Graphic Skills

Line Graphs

On the GED Science Test, you may answer questions in which you interpret data that is presented in a **line graph.**

- A **line graph** displays a change in one variable compared to another variable. Most line graphs show a change over time.
- To interpret a line graph, locate specific points on the graph and identify the **trend** that the graph represents.

Examine the graph. Choose the one best answer to the question.

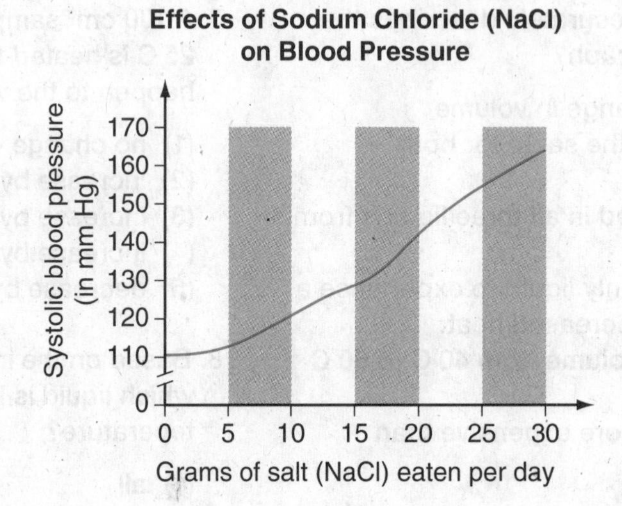

QUESTION: Which of the following is a conclusion supported by the graph?

(1) Salt has no effect on blood pressure.
(2) Eating 15 grams of salt per day is not healthy.
(3) Blood pressure tends to increase with increased daily salt intake.
(4) A healthy diet includes no salt.
(5) People with high blood pressure tend to dislike salt.

EXPLANATIONS

STEP 1 To answer this question, ask yourself:

- What do points on the line represent? blood pressure at different levels of NaCl
- What trend does the line show? a steadily increasing rise in blood pressure as salt (NaCl) consumption increases

STEP 2 Evaluate all of the answer choices and choose the best answer.

(1) No. As salt increases, blood pressure increases.
(2) No. There is no information about what constitutes a healthy diet.
(3) **Yes. According to the graph, there is a direct relationship between blood pressure and salt intake; as salt intake increases, so does blood pressure.**
(4) No. There is no information to indicate that salt should be avoided.
(5) No. There is no information about food preferences.

ANSWER: (3) Blood pressure tends to increase with increased daily salt intake.

Practice the Skill

Try these examples. Choose the one best answer to each question. Then check your answers and read the explanations.

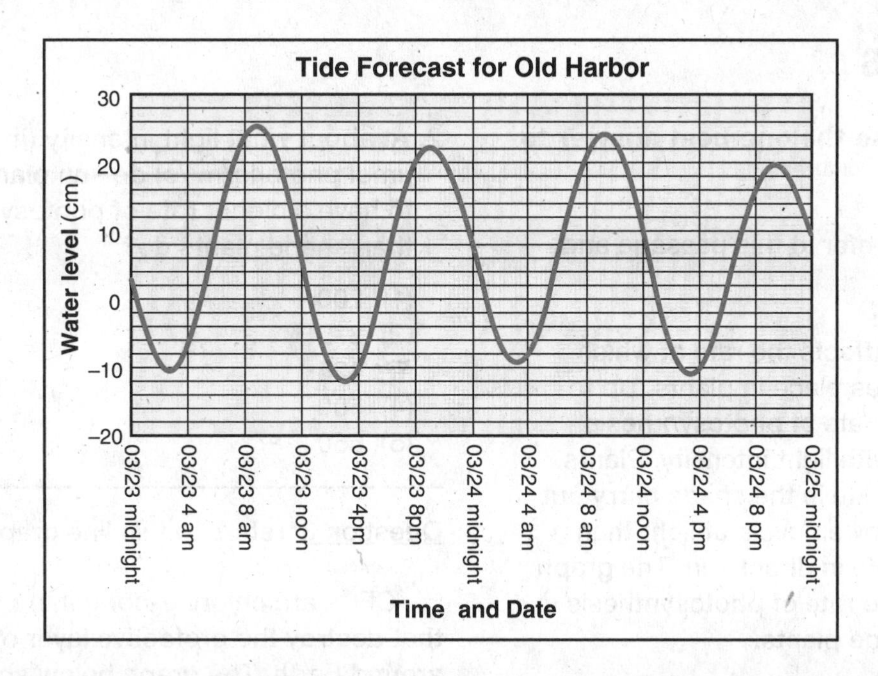

1. According to the graph, by how many centimeters does the water level change between the first high tide of March 23 and the following low tide?

(1) –12
(2) 26
(3) 31
(4) 34
(5) 38

HINT **Subtract the low-tide elevation from the high-tide elevation.**

2. What is the approximate time (in hours) between one high tide and the next?

(1) 3
(2) 4
(3) 6
(4) 12
(5) 24

HINT **Pay attention to the scale at the bottom of the graph.**

Answers and Explanations

1. (5) 38
Option (5) is correct: the difference between the high tide (26 cm) and the low tide (–12 cm) is 38 centimeters.

Values of –12 cm (option 1) and 26 cm (option 2) are the differences between the high and low tides and the zero level. Option 3—31 cm—is the difference between the second high tide of March 23 and the next low tide. Option 4—34 cm— is the difference between the second low tide and the second high tide.

2. (4) 12
Option (4) is correct because there are three vertical lines between each high tide, and each line represents 4 hours: 3 × 4 = 12.

Three hours (option 1) counts each section as 1 hour instead of 4. Four hours (option 2) is the scale of the graph. Six hours (option 3) is the approximate time between high tide and low tide, and 24 hours (option 5) is the period of time from one high tide to another and then another.

GED Practice

Line Graphs

Directions: Choose the one best answer to each question.

Questions 1 and 2 refer to this passage and line graph.

Light intensity affects the rate at which photosynthesis takes place in plants. Up to a certain point, the rate of photosynthesis tends to increase with light intensity. Plants that are adapted to life in the shade carry out photosynthesis at lower levels of light than plants adapted to life in direct sun. The graph below compares the rate of photosynthesis in sun plants and shade plants.

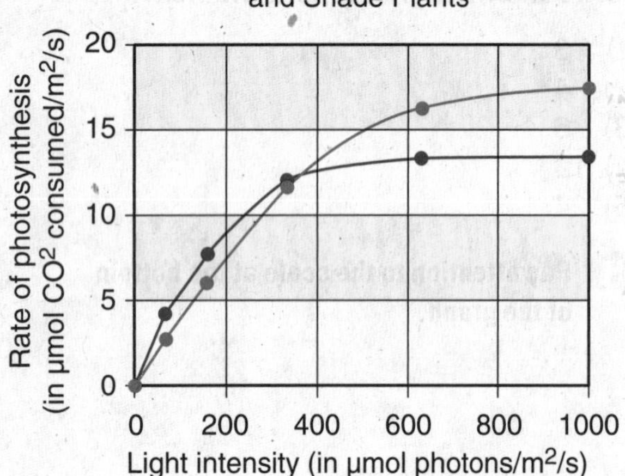

1. Where are shade plants likely to carry out photosynthesis at a higher rate than sun plants?
   (1) forest
   (2) desert
   (3) sunny meadow
   (4) window that gets direct sun
   (5) cave

2. At about what light intensity (in µmol photons/m²/s) do sun plants begin to have a higher rate of photosynthesis than shade plants do?
   (1) 200
   (2) 375
   (3) 425
   (4) 500
   (5) 650

Question 3 refers to this line graph.

CFCs are chlorine-containing chemicals that destroy the protective layer of ozone around Earth. The graph below shows the predicted effects of limiting the use of CFCs.

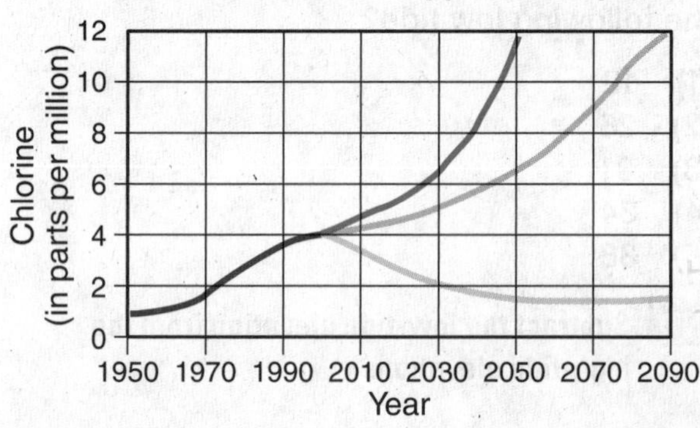

3. What is the predicted level of chlorine (in parts per million) in Earth's upper atmosphere in the year 2050 if CFCs are banned?
   (1) 0.8
   (2) 1.8
   (3) 2.2
   (4) 6.4
   (5) 11.8

Questions 4 through 7 refer to these line graphs.

The following graphs describe the relationships between speed, time, and distance for a train. The graph on the left below shows the relationship between the distance traveled and time. The graph on the right below shows the train's average speed during that time period.

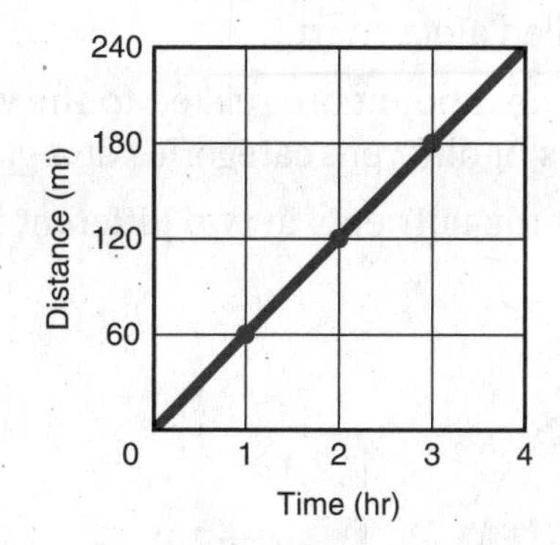

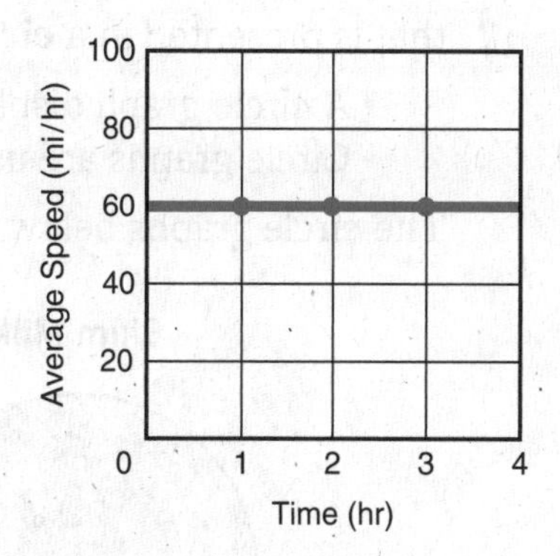

4. In the speed-time graph, the average speed is shown as a constant. What can be assumed about the speed of the train during the course of the four-hour trip?
   (1) The speed was determined by the distance.
   (2) The speed was constantly increasing.
   (3) The speed was constantly decreasing.
   (4) The actual speed of the train did not vary.
   (5) The average speed is not the same as the actual speed.

5. How far had the train traveled after three hours?
   (1) 4
   (2) 60
   (3) 120
   (4) 180
   (5) 240

6. What can be inferred about the relationship between the two graphs?
   (1) Distance is shown on one axis of each graph.
   (2) Distance and average speed are different names for the same measure.
   (3) Speed is shown on one axis of each graph.
   (4) The graphs show two aspects of the same trip.
   (5) Time is shown on a different axis of each graph.

7. Based on information in the graphs, which of the following conclusions is true about the four-hour trip?
   (1) As distance increases, the average speed also increases.
   (2) As time increases, the average speed also increases.
   (3) As time increases, distance also increases.
   (4) As distance increases, time decreases.
   (5) The average speed and time remain the same.

TIP

A line graph can show upward, downward, or static trends. To interpret this information, locate two points on the graph and draw a line between them to identify the trend of the graph.

Answers and explanations start on page 112.

Graphic Skills

Circle Graphs

On the GED Science Test, you will have to answer questions in which you interpret data that is presented in a **circle graph,** sometimes called a pie chart.

- A circle graph displays how different parts of an amount are related to the whole.
- Circle graphs are used to show **relative sizes** of different categories at a glance.

The circle graphs below present information about the nutrients in two different foods.

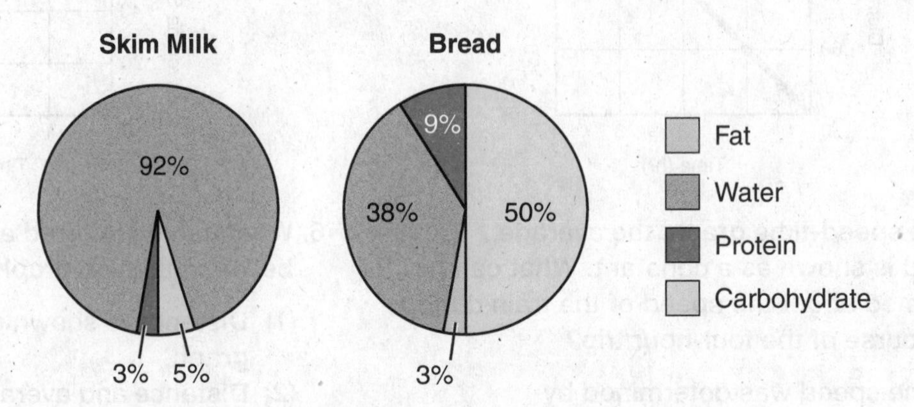

Examine the graphs. Choose the <u>one best answer</u> to the question.

QUESTION: The body gets energy from carbohydrates, protein, and fat. Based on this information, which of the following conclusions is supported by the graphs?

(1) Bread is a better source of energy than skim milk.
(2) Skim milk is a healthier food than whole milk.
(3) A healthy diet could consist of only bread.
(4) Skim milk is not nutritious because it is almost all water.
(5) Fats contain more energy than proteins.

EXPLANATIONS

STEP 1 To answer this question, ask yourself:

- What information do the graphs show? <u>relative amounts of nutrients in skim milk and bread</u>
- How can the two graphs be used together? <u>to compare the nutrients of the two foods</u>

STEP 2 Evaluate all of the answer choices and choose the <u>best</u> answer.

(1) Yes. Protein, carbohydrates, and fat provide energy, and 62% of bread contains these nutrients compared to only 8% in skim milk, so bread is a better source of energy.
(2) No. There is no information about whole milk.
(3) No. There is no information about the components of a healthy diet.
(4) No. The graph does not show information about the nutritional values of skim milk, only its nutrient composition.
(5) No. The graphs show relative amounts of fats and proteins in the foods, not the amount of energy that each provides.

ANSWER: (1) Bread is a better source of energy than skim milk.

Practice the Skill

Try these examples. Choose the one best answer to each question. Then check your answers and read the explanations.

Questions 1 and 2 refer to the following circle graph.

Landfall Patterns of Hurricanes Approaching the United States During a 5-Year Period

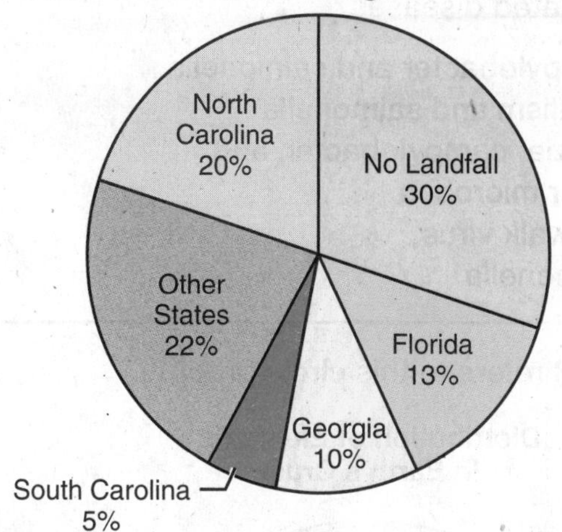

1. According to the graph, where did about one-third of all hurricanes make landfall?
   (1) Georgia and Florida
   (2) North Carolina and South Carolina
   (3) Florida and North Carolina
   (4) other states
   (5) countries other than the United States

HINT **Add the percents from the sections of the graph.**

2. Which of the following conclusions could be made from the data in the graph?
   (1) Most hurricanes made landfall in July.
   (2) Most hurricanes are classified as category 5.
   (3) More hurricanes made landfall in North Carolina than in Florida.
   (4) A state's coastline determines the number of hurricanes that make landfall there.
   (5) Ten hurricanes made landfall in Georgia.

HINT **Read the graph's title carefully. It often has important information.**

Answers and Explanations

1. (3) Florida and North Carolina
Option (3) is correct because the sum of the percents of hurricanes hitting land in Florida and North Carolina is 20% + 13% = 33%, or approximately one-third.

The sums of other combinations are 23% for Georgia and Florida (option 1) and 25% for North Carolina and South Carolina (option 2). Other states (option 4) total only 22%, which is closer to one-fifth, not one-third. There is no mention on the graph of hurricanes making landfall in other countries (option 5).

2. (3) More hurricanes made landfall in North Carolina than in Florida.
Option (3) is correct because the data gives the relative rates, or percents, for hurricanes that made landfall in each state.

There is no information on the graph about timing of hurricanes (option 1), intensity of hurricanes (option 2), or length of coastline (option 4). Although ten percent of hurricanes did make landfall in Georgia, the exact number of hurricanes (option 5) could only be calculated if the total number of hurricanes that occurred during the period had been provided.

GED Practice

Circle Graphs

Directions: Choose the one best answer to each question.

Questions 1 and 2 refer to this passage and circle graph.

Each year an estimated 79 million people in the U.S. become ill from food-related diseases due to food contaminated by microbes. About 14.3 million of the cases are tied to food-borne bacteria, parasites, and viruses. While foods are generally safe if handled and stored properly, they are not sterile. Poor hygiene contributes to many food-related outbreaks as does the improper cooking and care of food.

Causes of Food-Related Diseases

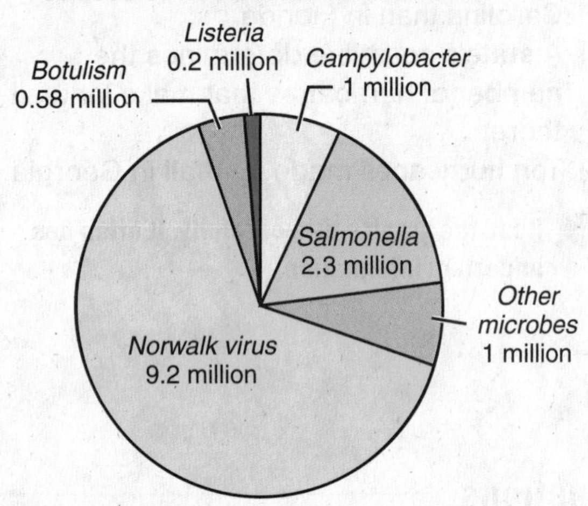

1. Based on the circle graph, which organism is the least likely cause of food-borne illnesses?
   (1) botulism
   (2) listeria
   (3) campylobacter
   (4) salmonella
   (5) Norwalk virus

2. What organism(s) is responsible for most food-related diseases?
   (1) campylobacter and salmonella
   (2) botulism and salmonella
   (3) listeria, campylobacter, and other microbes
   (4) Norwalk virus
   (5) salmonella

Question 3 refers to this circle graph.

Distribution of Elements in Earth's Crust

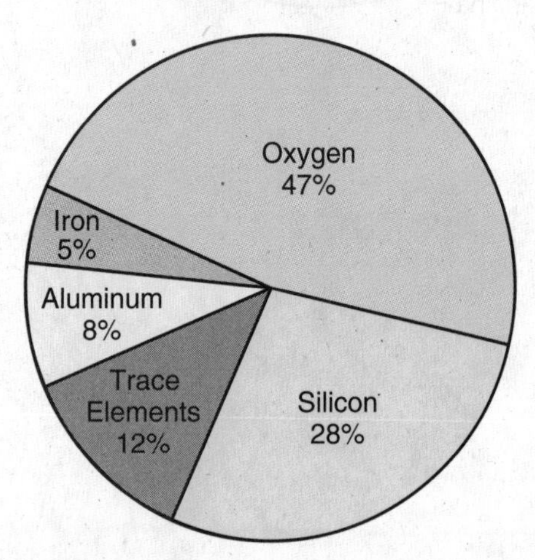

3. Silicates are compounds made of silicon and oxygen linked with other kinds of atoms. Based on information from the circle graph, why is a geologist most likely to find rocks on Earth's surface that are composed of silicates?
   (1) Silicon and oxygen are the two most abundant elements of Earth's crust.
   (2) Iron is too hard to form rocks.
   (3) Silicon and oxygen combine more easily than the other elements.
   (4) Silicon and oxygen are the only elements in Earth's crust.
   (5) Aluminum is found at a deeper depth than silicon.

Questions 4 through 7 refer to the circle graphs.

An unusual mortality event (UME) is an unexpected event that causes significant deaths in a marine mammal population.

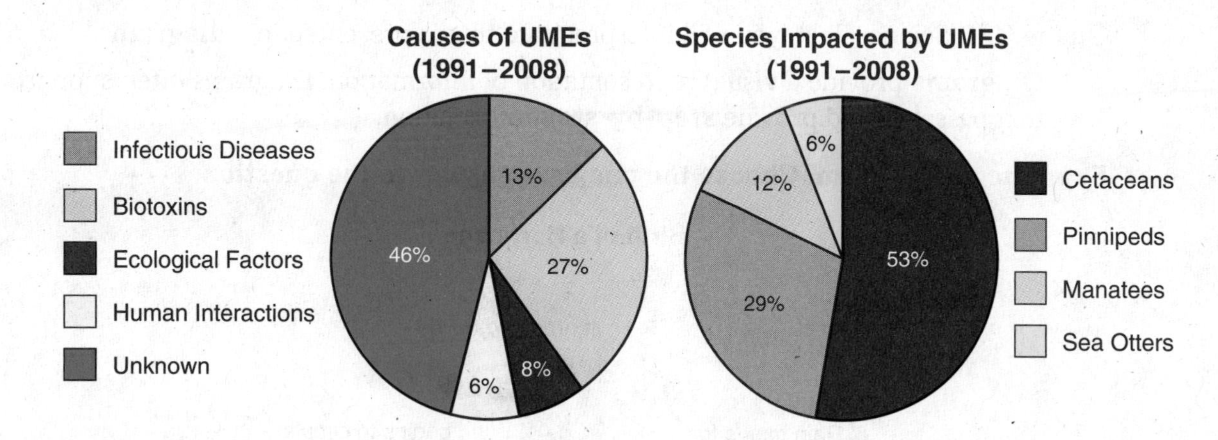

4. Which of the following would most likely be the cause of a UME?
   (1) oil spill
   (2) decline in seaweed
   (3) a new virus
   (4) increase in toxin-producing algae
   (5) decreased use of fishing boats

5. Which of these events was least likely to be a cause of an unexpected mortality event during the period 1991–2008?
   (1) infectious diseases
   (2) biotoxins
   (3) ecological factors
   (4) human interactions
   (5) all events are equally likely

6. Which of the following statements is true based on the information in the graph on the right?
   (1) Cetaceans have the largest total population.
   (2) Manatees swim faster than sea otters.
   (3) Pinnipeds are the most affected by UMEs.
   (4) Manatees are less affected by UMEs than cetaceans.
   (5) Cetaceans are less affected by UMEs than manatees.

7. Which animals had the fewest documented UMEs during the period?
   (1) cetaceans
   (2) sea otters
   (3) pinnipeds
   (4) manatees
   (5) manatees and sea otters combined

TIP

When you read a circle graph, ask yourself the following questions:
- **What does the entire circle represent?**
- **What does each sector of the circle represent?**
- **How do the percents of each of the sectors compare?**

Answers and explanations start on page 112.

Graphic Skills

Diagrams

On the GED Science Test, you will interpret data that is presented in a **diagram.**

- **Diagrams** provide a visual representation of information. Diagrams often support text passages and provide **step-by-step** information.

Examine the diagram. Choose the one best answer to the question.

Birth of a Hurricane

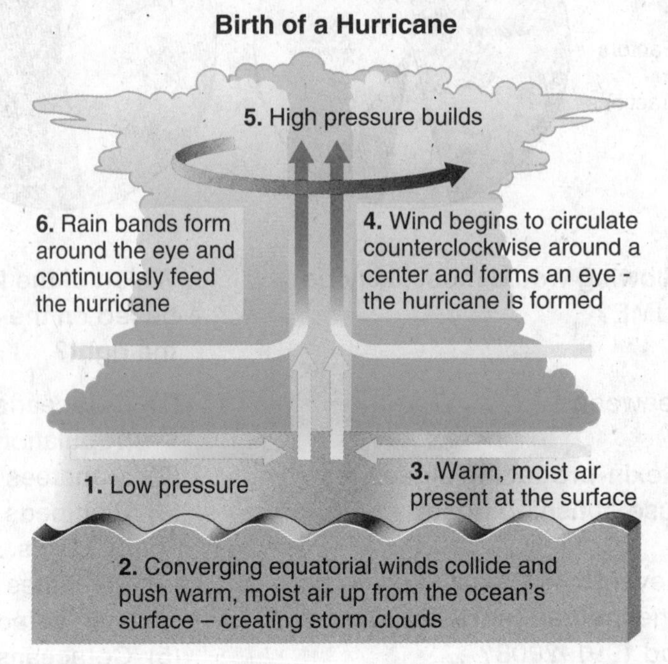

QUESTION: Which condition is not necessary for hurricane formation?

(1) low surface pressure
(2) formation of rain bands
(3) converging equatorial winds
(4) warm, moist air
(5) counterclockwise wind circulation

EXPLANATIONS

STEP 1 To answer this question, ask yourself:

- What information is presented in the diagram? hurricane formation
- What information do I need? the events that happen to form a hurricane

STEP 2 Evaluate all of the answer choices and choose the best answer.

(1) No. Low pressure is a necessary condition as shown in step 1.
(2) **Yes. Rain bands are a condition needed to fuel a hurricane, but are not necessary for the initial formation. They form around the eye of the hurricane after it has already formed.**
(3) No. As shown in step 2, converging equatorial winds are necessary.
(4) No. According to step 3, warm, moist air must be present at the water surface.
(5) No. Winds must circulate counterclockwise to form the eye of the hurricane, so this is a necessary condition.

ANSWER: (2) formation of rain bands

Practice the Skill

Try these examples. Choose the one best answer to each question. Then check your answers and read the explanations.

Questions 1 and 2 refer to this diagram.

Photosynthesis and Cellular Respiration in Plants

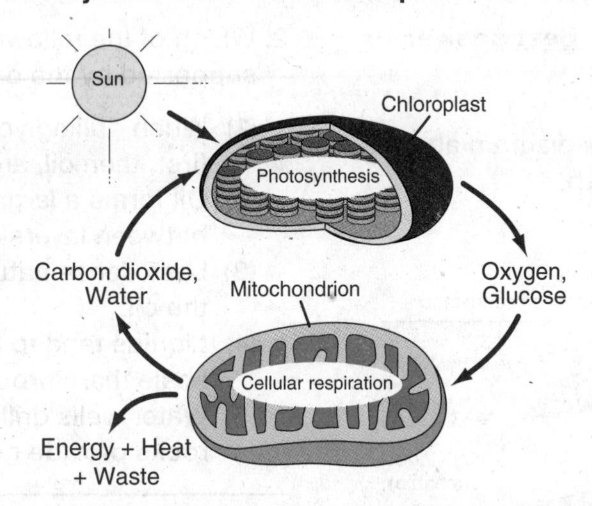

1. Based on the diagram, what are the ingredients needed for photosynthesis?
   (1) oxygen and glucose
   (2) light energy, carbon dioxide, and water
   (3) energy, heat, and waste
   (4) carbon dioxide and oxygen
   (5) light energy

HINT **Use the arrows to determine the relationships between the different parts of the diagram.**

2. If only plants carry out photosynthesis, what can you conclude about the chemical results of cellular respiration in animals?

   They are

   (1) used to make the animal's food
   (2) produced in trace amounts
   (3) excreted from the body as waste
   (4) used to carry out cellular respiration
   (5) converted to light energy

HINT **What would happen to the chemical products produced by animals?**

Answers and Explanations

1. (2) light energy, carbon dioxide, and water
Option (2) is correct because according to the diagram, energy from the sun, carbon dioxide, and water enter the chloroplast, which is where photosynthesis occurs.

Option (1) is incorrect because glucose and oxygen leave the chloroplast after photosynthesis. Option (3) is incorrect because energy, heat, and waste are the results of cellular respiration. Options (4) and (5) are incorrect because they are only partial lists, not complete.

2. (3) excreted from the body as waste
Option (3) is correct. Carbon dioxide and water are the chemical results of cellular respiration as shown in the diagram. Since animals do not carry out photosynthesis, their chemical products would leave their bodies as waste.

Animals do not make their own food, so option (1) is incorrect. The amount of materials involved is not presented (option 2). Option (4) contradicts the diagram. Light energy is an input needed for photosynthesis, which is not a process carried out by animal cells (option 5).

GED Practice

Diagrams

<u>Directions</u>: Choose the <u>one best answer</u> to each question.

<u>Questions 1 and 2</u> refer to this diagram about a rock structure called an oil trap.

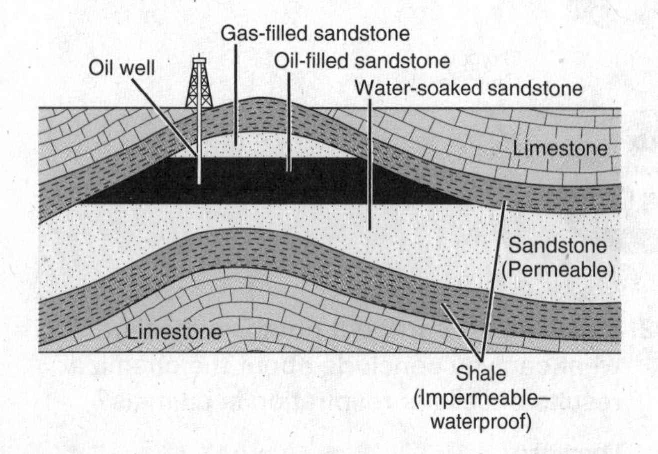

1. The caprock is the impermeable rock that stops oil from reaching the surface. What is the cap rock in the diagram?
   (1) shale above the sandstone
   (2) sandstone
   (3) limestone above the shale
   (4) shale beneath the sandstone
   (5) limestone below the shale

2. Which of the following statements is supported by the diagram?
   (1) When drilling you are likely to hit gas first, then oil, and then water.
   (2) Oil forms a large underground lake between layers of rock.
   (3) Less dense natural gas collects below the oil.
   (4) Liquids tend to flow more easily through shale than through sandstone.
   (5) Water wells drilled into impermeable rocks provide good water sources.

<u>Question 3</u> refers to this diagram.

Isotopes of Hydrogen

Protium　Deuterium　Tritium

+ = proton　n = neutron　– = electron

3. Which statement is supported by the information in this diagram?
   (1) Hydrogen is an element.
   (2) Deuterium is not a type of hydrogen.
   (3) All hydrogen atoms have only one neutron.
   (4) Hydrogen is the only element that has different isotopes.
   (5) Isotopes have the same number of protons but different numbers of neutrons.

TIP

Always read the captions and labels on drawings and diagrams. Both captions and labels may provide important details needed to answer questions. Labels can help you understand the parts or the steps included in a diagram.

Questions 4 through 7 refer to this passage and diagram.

The word *respiration* refers both to the process of breathing and to the biochemical changes by which cells obtain and store energy from food. In an animal that breathes air, external respiration takes place in the lungs. The relationship between the two types of respiration is shown in this flowchart.

External and Internal Respiration

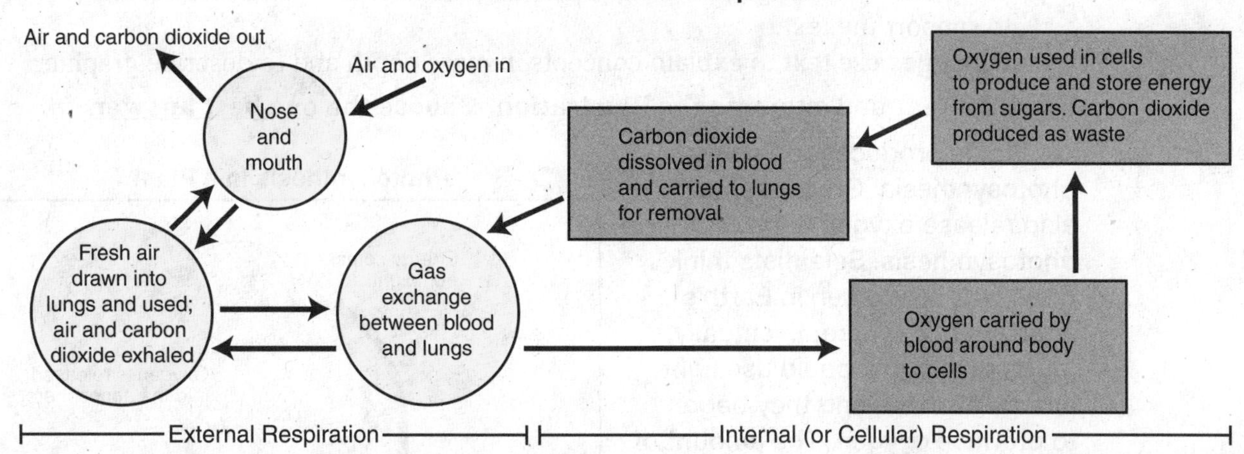

4. What occurs first in external respiration?
   (1) The lungs breathe in air and oxygen through the nose and mouth.
   (2) The blood carries oxygen to body cells.
   (3) Cells produce carbon dioxide.
   (4) Carbon dioxide is carried to the lungs.
   (5) Gas is exchanged between blood and lungs.

5. Which organ described below is most similar in function to the lungs described in the diagram?
   (1) lateral line, which picks up vibrations from the water
   (2) scales, which cover the surface of a fish's body
   (3) gills, which exchange carbon dioxide for oxygen in water
   (4) kidneys, which regulate the amount of water in the body
   (5) heart, which pumps blood through the body

6. What is a likely effect of an increased rate of cellular respiration?
   (1) a decrease in the amount of carbon dioxide produced by the body
   (2) an increase in the rate at which air and oxygen are inhaled and exhaled
   (3) a decrease in the rate at which the heart pumps blood throughout the body
   (4) an increase in the amount of oxygen produced by cellular respiration
   (5) a decrease in the amount of energy produced by the body

7. If an athlete's lung and heart capacity increases, what else is most likely to increase?
   (1) muscle mass
   (2) strength
   (3) energy
   (4) coordination
   (5) balance

Answers and explanations start on page 113.

Combine Information from Multiple Sources

Combine Text and Graphics

On the GED Science Test, you will have to answer questions in which you use a combination of **text information** and **graphic information.**

- **Graphics**—tables, graphs, diagrams, and illustrations—provide visual information to support the text.
- **Passages** use text to explain concepts in more depth and to describe graphics.

Read the text and examine the illustration. Choose the one best answer.

Plants produce food using photosynthesis. Green plants also release oxygen during photosynthesis. Scientists think there was no oxygen in Earth's ancient atmosphere. Eventually, green plants that could use light energy evolved, and they began to produce oxygen. The amount of oxygen increased as these plants grew. Today oxygen accounts for about one-fifth of the amount of gas in the atmosphere.

Photosynthesis in a Plant

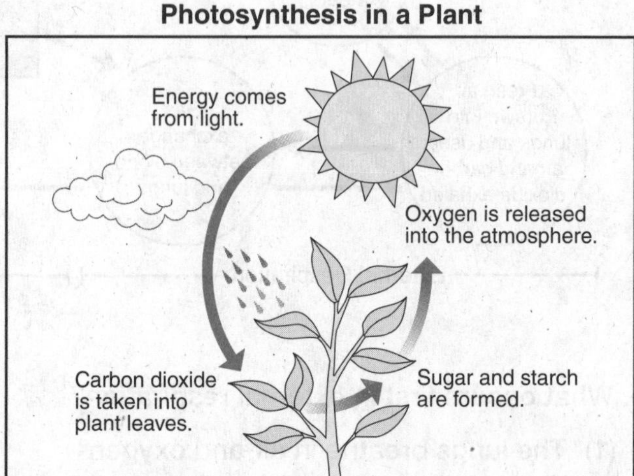

QUESTION: Which conclusion is based on the diagram and the passage?

(1) Oxygen levels would be much lower without light energy.
(2) Photosynthesis will eventually consume all the oxygen on Earth.
(3) Photosynthesis releases carbon dioxide.
(4) Photosynthesis only occurs with natural sunlight, not artificial light.
(5) Only plants on land make food using photosynthesis.

EXPLANATIONS

STEP 1 To answer this question, ask yourself:

- What information is presented? evolution of plants and photosynthesis
- What do I know about photosynthesis? produces sugar, starch, and oxygen

STEP 2 Evaluate all of the answer choices and choose the best answer.

(1) **Yes. The diagram and passage indicate that light energy is needed for photosynthesis, and the passage and diagram both indicate that photosynthesis produces oxygen.**
(2) No. The passage and diagram both indicate that oxygen is a product of photosynthesis, not an ingredient.
(3) No. The passage and the diagram indicate that carbon dioxide is used by, not released by, photosynthesis.
(4) No. Although the diagram shows that sunlight powers the process, the passage uses the phrase "light energy" and does not mention sunlight specifically.
(5) No. The passage and diagram imply that all plants use photosynthesis.

ANSWER: (1) Oxygen levels would be much lower without light energy.

Practice the Skill

Try these examples. Choose the one best answer to each question. Then check your answers and read the explanations.

Questions 1 and 2 refer to this diagram and passage.

Winds result from uneven heating of the atmosphere. As air becomes heated, it expands and rises. This creates an area of low pressure near Earth's surface. Air molecules tend to move from areas of high pressure to areas of low pressure. Cooler air generally moves toward this low-pressure area. This moving cold air creates wind. Uneven heating of air above land and water causes two kinds of local winds, called breezes.

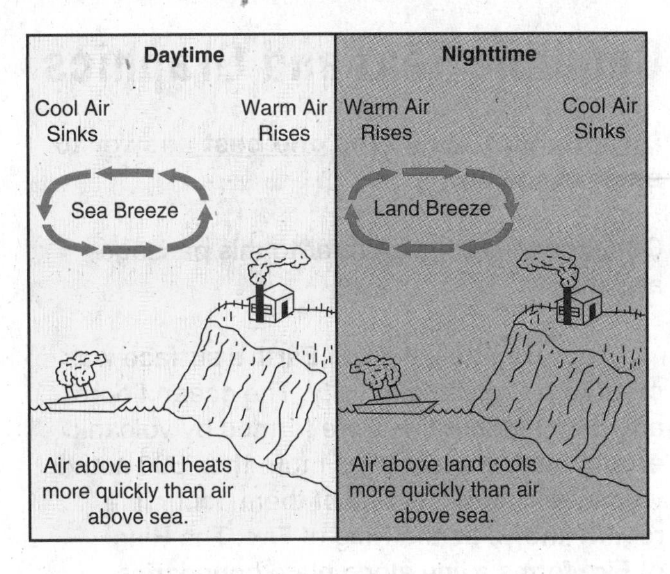

1. What happens when air above the land becomes warmer than air above the sea?
   (1) Warmer air from the sea replaces rising cool air, forming a land breeze.
   (2) The warmer and cooler air masses come together and form storms.
   (3) Air above the land sinks while air above the sea rises.
   (4) Air pressure over the sea decreases.
   (5) Warm air rises and is replaced by cooler air, forming a sea breeze.

HINT **Apply what you learned about winds in the passage to determine how temperature affects the air.**

2. Why are land breezes more likely to form during the night?
   (1) Water does not cool as quickly as soil and rock.
   (2) Air tends to rise over land during the night.
   (3) Air pressure decreases over land at night.
   (4) Air over land cools and rises faster than air over water.
   (5) The warm air moves toward cooler air.

HINT **Use all of the available graphic information, including labels and descriptions.**

Answers and Explanations

1. (5) Warm air rises and is replaced by cooler air, forming a sea breeze.
Option (5) is correct because the passage states that cooler air replaces warm air, and the diagram illustrates this as a sea breeze.

According to the passage, warm air rises (option 1). Storm formation (option 2) is not discussed in either the diagram or the passage. According to the diagram, air above land rises during the day (option 3). The diagram indicates that cool air sinks over the ocean, so pressure increases, not decreases (option 4).

2. (1) Water does not cool as quickly as soil and rock.
Option (1) is correct. The diagram shows that water cools more slowly than land, thus the air above water does not cool as quickly as air above land.

Option (2) contradicts the diagram, which shows warm air rising over land during the day, not the night. Air pressure over land increases at night (option 3). Cooler air does not rise (option 4). The passage states that cooler air moves toward the low pressure created by warmer air (option 5).

Combine Text and Graphics

Directions: Choose the one best answer to each question.

Questions 1 through 4 refer to this passage and map.

Approximately 80% of Earth's surface was created by volcanic activity. The ocean floors and many mountains were formed by volcanic eruptions. Our planet has more than 500 active volcanoes, and over half of them occur in a region known as the Ring of Fire. The Ring of Fire forms a line along plate boundaries, extending from Indonesia, along the eastern coast of Asia, and the western coasts of North and South America.

Locations of Convergent and Divergent Plates

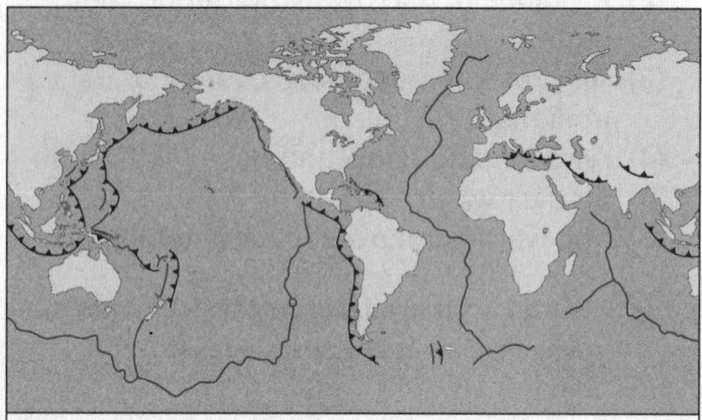

KEY

Divergent plate boundaries—Where new crust is generated as the plates pull away from each other.

Convergent plate boundaries—Where crust is consumed in Earth's interior as one plate dives under another.

1. Based on the information in the map and passage, which region of the United States is most likely to experience volcanic eruptions?
   (1) the northeast coast
   (2) the southeast coast
   (3) the central region
   (4) the western coast
   (5) equal likelihood in all regions

2. Based on the information presented, what is volcanic activity likely to do?
   (1) cease within the next 100 years
   (2) decline in the Ring of Fire and increase elsewhere
   (3) continue to occur along plate boundaries
   (4) cause the destruction of major cities
   (5) depend on the rate of global warming and climate change

3. Based on the information in the passage and diagram, what conclusion can you make about the line on the map that extends through the Atlantic Ocean?
   (1) The line shows a section of the Ring of Fire.
   (2) Volcanoes do not occur there because it is not part of the Ring of Fire.
   (3) New crust is generated on the ocean floor by volcanoes along the line.
   (4) A new continent will eventually form along the line due to volcanic activity.
   (5) One part of the crust is sliding over another part in the middle of the Atlantic Ocean.

4. Why do more volcanoes occur near the edges of continents than in their central regions?
   (1) Volcanoes are generally caused by the weight of ocean water on the crust.
   (2) There are few plate boundaries within the continents.
   (3) Volcanoes only occur at the Ring of Fire.
   (4) The crust is thicker in the middle of a continent.
   (5) Volcanoes only occur at convergent plate boundaries.

Questions 5 through 8 refer to this passage and diagram.

Some isotopes are called radioactive because they are unstable and decay. When they decay, their nuclei break apart and give off energy. The product of this decay can be another unstable isotope or a stable isotope. The time it takes for half the atoms of an unstable isotope to decay to an end product is called the half-life. Every radioactive isotope (or radioisotope) has its own constant rate of decay and half-life. Half-lives range from less than a second to billions of years. Thorium-234 has a relatively short half-life. A graphic representation of the decay of a 16-gram sample of thorium-234 is shown below.

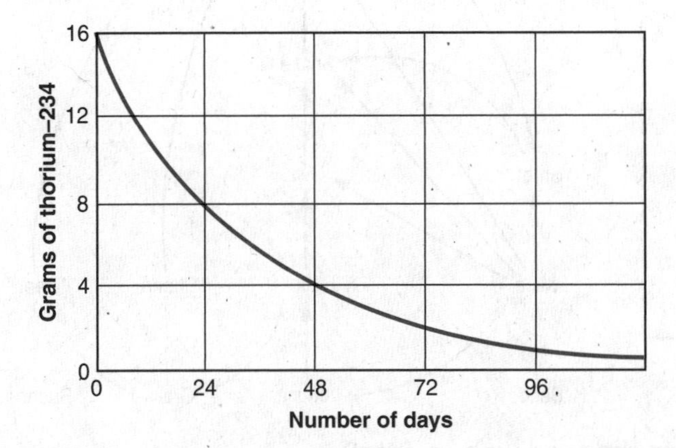

5. Which of the following statements supports the conclusion that a measurement of the amount of a radioisotope remaining in a rock can be used to determine the time elapsed since the rock formed?

   (1) Every radioisotope has its own constant rate of decay and half-life.
   (2) Thorium-234 has a relatively short half-life.
   (3) Some isotopes are called radioactive because they are unstable and decay.
   (4) After 48 days, 16 grams of thorium-234 decays to 4 grams.
   (5) The half-life is the time it takes for half the atoms of an unstable element to decay.

6. Which of the following statements supports the conclusion that the half-life of thorium-234 is 24 days?

   (1) After 24 days, about 4 grams of thorium-234 have decayed.
   (2) After 48 days, about 8 grams of thorium-234 have decayed.
   (3) Every radioisotope has its own constant rate of decay and half-life.
   (4) On the 24th day, only 8 of the original 16 grams of thorium-234 remain.
   (5) Some half-lives are less than a second.

7. After three half-lives, how many grams of thorium-234 remains?

   (1) 16
   (2) 8
   (3) 6
   (4) 4
   (5) 2

8. Uranium-238 in a rock sample decays to form lead-205. The half-life of uranium-238 is 4.5 billion years. What will be the ratio of uranium-238 to end product (lead-205) in the rock sample after 4.5 billion years?

   (1) 1:0
   (2) 1:1
   (3) 1:2
   (4) 2:1
   (5) 1:4

TIP

Look in your local newspaper's health and science sections, and read the articles that include a graphic. Test your information-gathering skills by asking yourself these questions: Would the article be as informative without the graphic? Could the graphic convey all the necessary information without the article?

Answers and explanations start on page 113.

Combine Information from Multiple Sources

Combine Information from Graphics

On the GED Science Test, you will have to answer questions in which you use a combination of **graphic information** from two or more sources.

- **Tables, graphs, diagrams,** and **illustrations** provide information visually.
- Multiple graphics can be used to relate complex or large amounts of information in smaller parts that are easier to understand.

The two diagrams below present information about relationships of organisms.

Examine the diagrams. Choose the one best answer to the question.

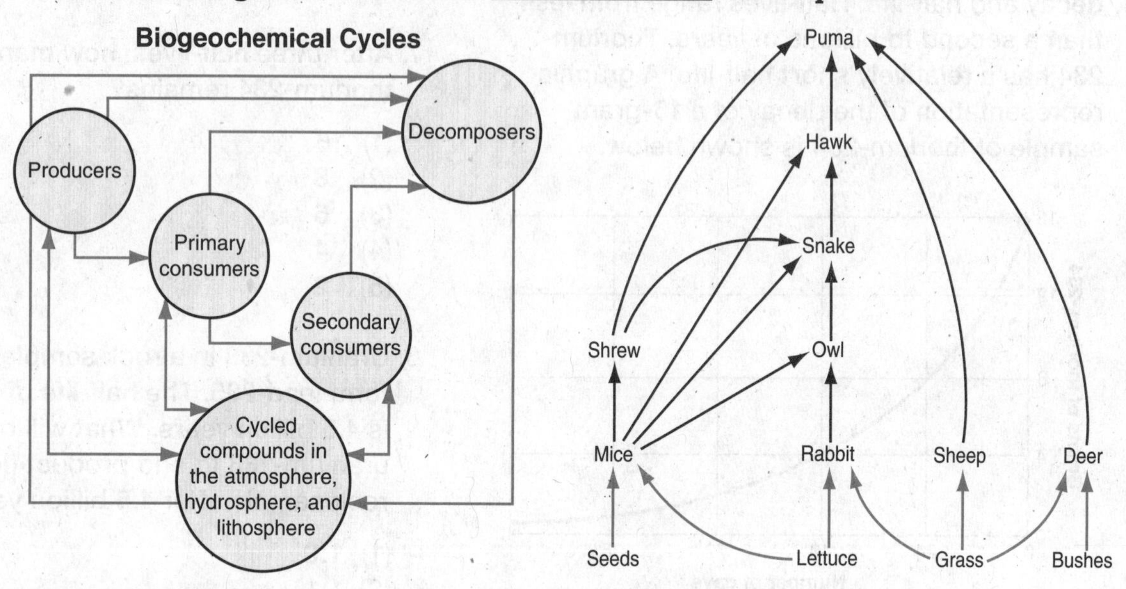

QUESTION: What is the role of snakes in the food web at right?

(1) producer
(2) primary consumer
(3) secondary consumer
(4) decomposer
(5) both primary and secondary consumer

EXPLANATIONS

STEP 1 To answer this question, ask yourself:

- What information is presented? relationships among organisms through a food web and the roles the organisms play in the biogeochemical cycle
- What do I need to do to answer the question? relate the information about the organisms from one diagram with information in the other diagram

STEP 2 Evaluate all of the answer choices and choose the best answer.

(1) No. Producers gain energy from compounds in the environment.
(2) No. Primary consumers obtain energy from producers such as plants.
(3) **Yes. Secondary consumers obtain energy from other consumers.**
(4) No. Decomposers recycle materials back into the environment.
(5) No. The snake only eats consumers, not producers, which means it is only a secondary consumer.

ANSWER: (3) secondary consumer

Practice the Skill

Try these examples. Choose the one best answer to each question. Then check your answers and read the explanations.

Questions 1 and 2 refer to these diagrams.

Diagram of an Animal Cell

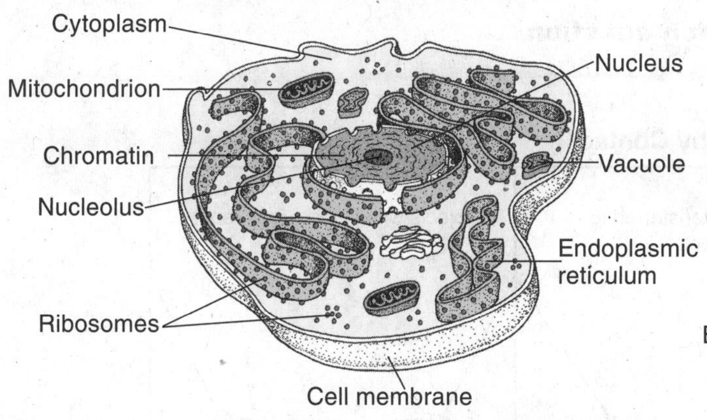

Diagram of a Plant Cell

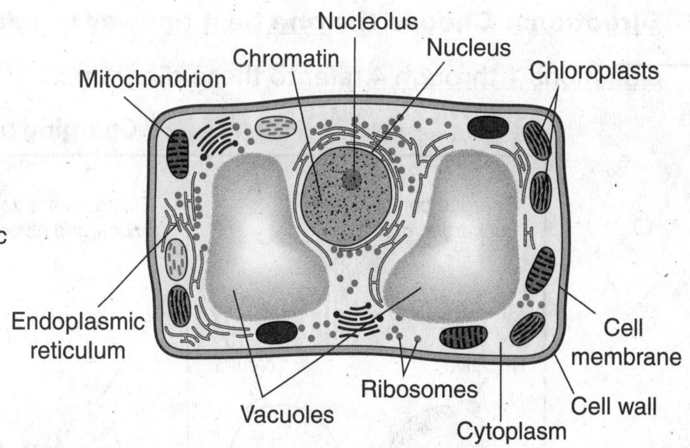

1. Plant cells tend to be very rigid compared to animal cells. Which cell structure is most likely to be responsible for this difference?

(1) cell membrane
(2) cell wall
(3) chloroplasts
(4) ribosomes
(5) nucleus

HINT **Look for differences between the two types of cells shown in the diagrams.**

2. Animals are consumers, and plants are producers. What difference in the diagrams is most likely the key part of the difference in these roles in an ecosystem?

(1) The cells have different shapes.
(2) Animal cells have mitochondria.
(3) Plant cells have chloroplasts.
(4) The shape of the vacuoles is different.
(5) Plant cells have vacuoles.

HINT **What types of structures occur only in one type of cell?**

Answers and Explanations

1. (2) cell wall
Option (2) is correct because the cell wall is the structural part of the two cells that is different. It surrounds and contains the cell, so it is most likely to affect the rigidity.

A comparison of the two diagrams shows that both types of cells have a cell membrane (option 1) and a nucleus (option 5), so these structures are not likely to account for the difference. Both cells have ribosomes (option 4), so this isn't a likely difference. Only plant cells have chloroplasts (option 3), but they are small structures inside the cell, so they are less likely to cause the cell to be rigid.

2. (3) Plant cells have chloroplasts.
Option (3) is correct because chloroplasts are only found in plant cells. They are involved in photosynthesis, which is the process by which plants produce food.

The shapes of the cells (option 1) do not account for the different functions of the cells. Both plant and animal cells have mitochondria (option 2) and vacuoles (option 5), so that is not a difference. The difference in shape of the vacuoles (option 4) appears to be due to a difference of perspective of the two diagrams.

GED Practice

Combine Information from Graphics

Directions: Choose the one best answer to each question

Questions 1 through 4 refer to these diagrams.

Charging by Contact

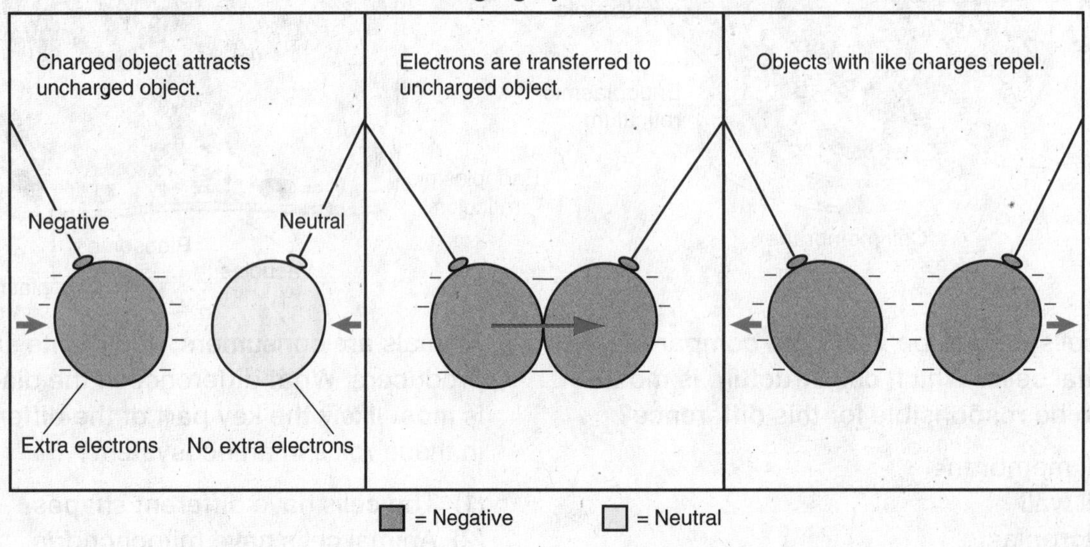

1. Which of the following is assumed the reader already knows?
   (1) that all balloons repel each other
   (2) that neutral objects have no extra electrons
   (3) how uncharged objects get negative charges
   (4) how uncharged objects get positive charges
   (5) what a negative charge is

2. What is the purpose of the series of diagrams?
   (1) to show that all materials, including balloons, contain electrons
   (2) to demonstrate that electrons can be transferred between objects that touch one another
   (3) to show the color change that occurs when an object becomes charged
   (4) to illustrate properties of balloons
   (5) to provide a warning about the danger of electric shock from touching a charged object

3. When a negatively charged table tennis ball is placed near a neutral table tennis ball, which of the following will most likely happen?
   (1) The balls will move apart.
   (2) The balls will not move.
   (3) A positive charge will be transferred to the neutral ball.
   (4) The negatively charged ball will lose its charge.
   (5) The balls will move together, touch, and then move apart.

4. Under the conditions shown in the diagram, which of the following scenarios results from charging by contact?
   (1) two positively charged objects
   (2) two neutral objects
   (3) two negatively charged objects
   (4) one negatively charged and one positively charged object
   (5) one negatively charged and one neutral object

Question 5 refers to this diagram and table.

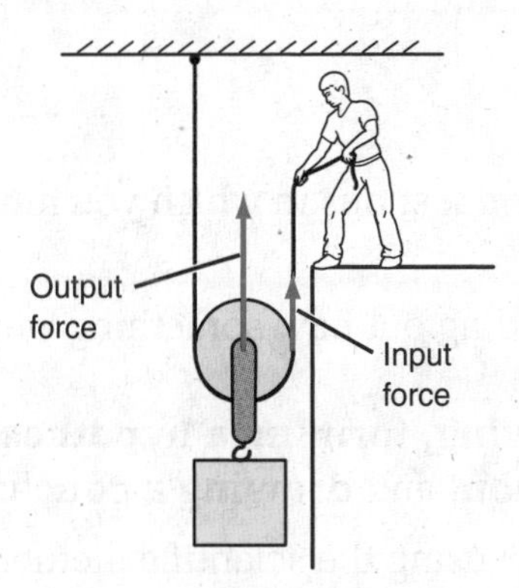

Note: The length of the arrows is relative to the strength of the force.

Change in Force for a Simple Pulley	
Input force (N)	**Output force (N)**
100	200
300	600
500	1,000

5. Based on the information in the diagram and table, if the man pulls up on the rope with a force of 200N, with what force will the box be lifted?
   (1) 100
   (2) 200
   (3) 400
   (4) 600
   (5) 800

Questions 6 and 7 refer to these diagrams.

Ear Anatomy

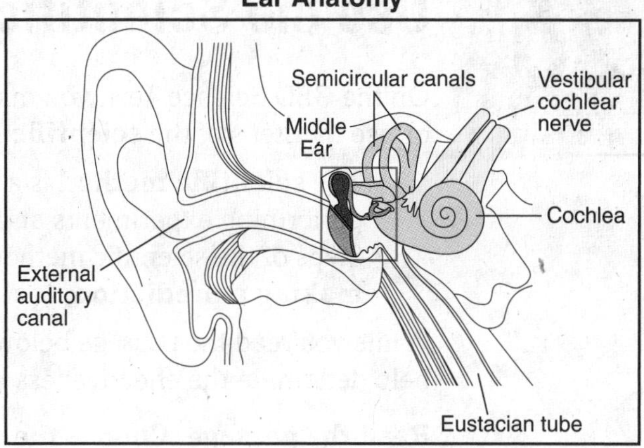

Middle Ear

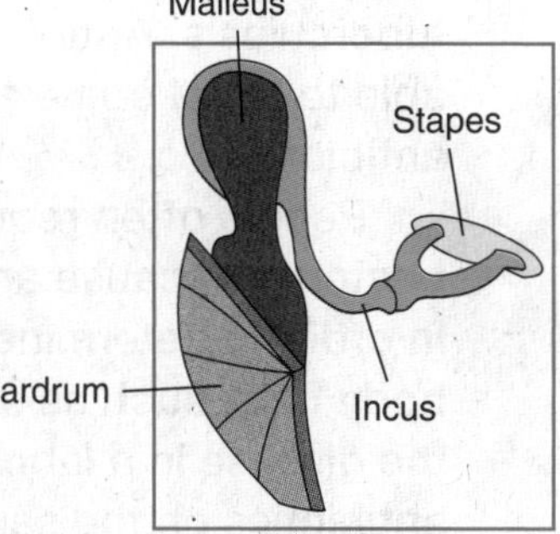

6. What structure of the middle ear transfers vibrations to the cochlea by direct contact?
   (1) eardrum
   (2) incus
   (3) malleus
   (4) stapes
   (5) vestibular cochlear nerve

7. Conductive hearing loss results from damage to the inner ear, preventing sound vibrations from reaching the inner ear. Which of the following problems would most likely cause a conductive hearing loss?
   (1) a buildup of fluid in the auditory canal
   (2) damage to the auditory nerve
   (3) a middle ear infection
   (4) damage to the eustacian tube
   (5) an injury to the outer ear

TIP

When there are two diagrams presented, look for the main concepts that link one diagram to the other.

Answers and explanations start on page 114.

Use the Scientific Method

On the GED Science Test, you may have to answer questions in which you must identify or use the steps of the **scientific method.**

- The **scientific method** is a procedure for finding out how something works by performing experiments and analyzing data.
- Steps of the scientific method include: **observing, forming a hypothesis, making a prediction, testing the prediction,** and **drawing a conclusion.**

While you read the passage below, think about how using the scientific method could help determine the effectiveness of an antibiotic.

Read the passage. Choose the one best answer to the question.

Penicillin and other antibiotics cure bacterial infections by killing bacterial cells or stopping their reproduction. Antibiotics have been used to cure pneumonia and tuberculosis. Antibiotics do not always work. Some disease-causing bacteria are able to resist some antibiotics but not others, so it is necessary to choose the right antibiotic.

People often request antibiotics for colds or flu, which are caused by viruses, not bacteria. Because antibiotics do not affect viruses, they do not cure these diseases. In order to determine the best way to treat a disease, doctors take a sample of a body fluid, such as saliva, from the patient in order to grow the organisms causing the disease in a laboratory. They can then test the effectiveness of different antibiotics on the sample rather than the patient.

QUESTION: A sample taken from a patient showed the disease organisms died when exposed to certain antibiotics but not when exposed to others. What conclusion could be made about the infection based on results of the experiment?

(1) A virus caused the disease.
(2) A type of bacteria did not cause the disease.
(3) A combination of viruses and bacteria were responsible for the disease.
(4) There is no effective treatment for the disease because some antibiotics do not work.
(5) It should be possible to cure the disease with antibiotics.

EXPLANATIONS

STEP 1 To answer this question, ask yourself:

- What was the experiment? exposing the organism to different antibiotics
- What was observed? Some antibiotics killed the organisms.

STEP 2 Evaluate all of the answer choices and choose the best answer.

(1) No. Antibiotics do not affect viruses, and since some antibiotics did kill the viruses, the organism must be a type of bacteria.
(2) No. Some antibiotics worked, so a bacteria probably did cause the disease.
(3) No. It can't be a combination because viruses can't be treated with antibiotics.
(4) No. Though some antibiotics didn't work, some did, which means there is most likely an effective treatment.
(5) **Yes. If the right antibiotic is used, it should cure the disease.**

ANSWER: (5) It should be possible to cure the disease with antibiotics.

Practice the Skill

Try these examples. Choose the one best answer to each question. Then check your answers and read the explanations.

Questions 1 and 2 refer to the following passage and line graph.

Photovoltaic (PV) cells generate electrical energy from sunlight. Because PV cells are expensive to manufacture and because only a small percentage of the sunlight that strikes a PV cell is converted to energy, they have been an expensive source of electrical energy. Researchers have improved the cells in recent years and more improvements are expected. In 2007, the U.S. Department of Energy released information showing the expected cost of PV energy. PV cells are expected to eventually be competitive with coal generators, whose energy costs about 4 cents per kilowatt hour.

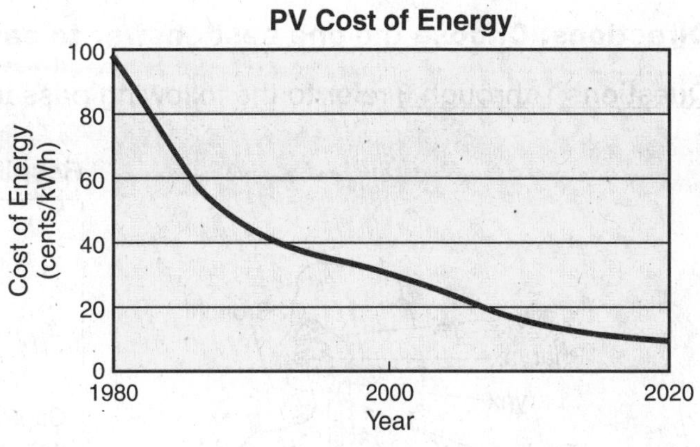

Source: U. S. Dept of Energy

1. According to the graph, the cost of PV energy in 2020 will be about 10 cents/kWh. Which term best describes this statement?
   (1) observation
   (2) prediction
   (3) variable
   (4) hypothesis
   (5) theory

HINT **What is another term for looking at data in the future?**

2. Based on the information in the passage, what research projects could be used by scientists to reduce the cost of PV energy?
   (1) building larger PV cells
   (2) increasing the efficiency of PV cells to convert light to energy
   (3) increasing the cost of energy produced by burning coal
   (4) finding new sources of energy to replace fossil fuels
   (5) reducing energy consumption

HINT **What is the limitation on the production of energy using PV cells?**

Answers and Explanations

1. (2) prediction
Option (2) is correct because a prediction is a statement about likely future events based on observations or experimental results.

An observation (option 1) is information collected by using your senses. A variable (option 3) is a factor that changes during an experiment. A hypothesis (option 4) is a possible explanation for an event or process. A theory is an explanation based on a number of experiments (option 5). These options do not correctly describe the information on the graph.

2. (2) increasing the efficiency of PV cells to convert light to energy
Option (2) is correct because the passage states that the limitation on production of PV energy is the efficiency of the cells.

There is no indication the size of PV cells (option 1) affects the cost of energy production. Increases in the cost of coal (option 3) may make PV energy more competitive but would not decrease its cost. Finding new energy sources (option 4) and reducing consumption (option 5) will increase the effectiveness of energy use but not decrease costs of PV cell production.

GED Practice

Use the Scientific Method

Directions: Choose the one best answer to each question

Questions 1 through 3 refer to the following passage and diagram.

Respiration

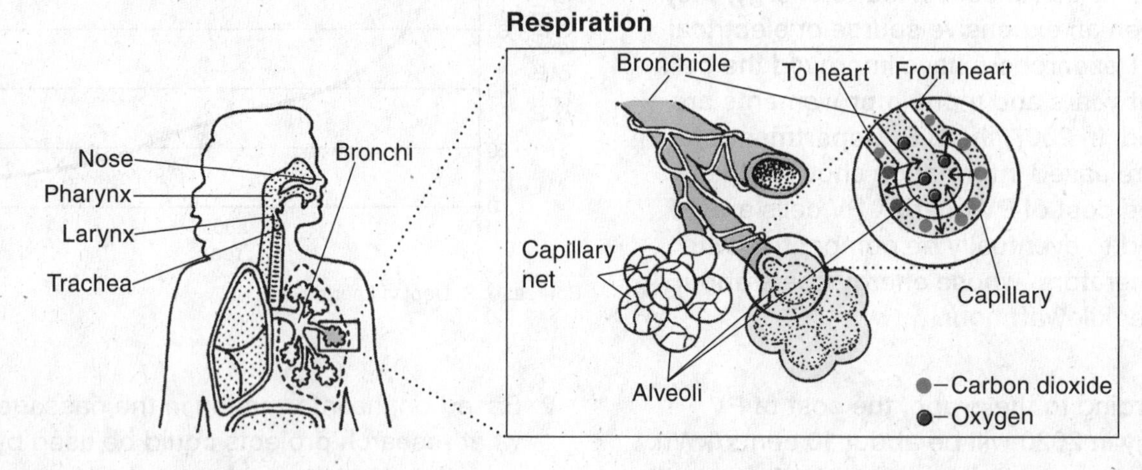

Medical researchers have found that air passes through the nose and trachea, which branches into two bronchi. In the lungs, these tubes branch again and again to form smaller and smaller bronchioles. At the ends of the smallest bronchioles are clusters of air sacs called alveoli. These have many tiny blood vessels where waste carbon dioxide in the blood is exchanged for oxygen from the air.

Researchers have observed patients to determine how air comes in and out of the body. From their observations, they hypothesized that air is sucked into the lungs when the diaphragm, a large muscle below the lungs, contracts. When the diaphragm relaxes, air is pushed out. Breathing is regulated by the respiratory control center of the brain.

1. Which conclusion can you draw from the information presented?

(1) Carbon dioxide and oxygen are exchanged in the lungs.
(2) People with bronchitis cough a lot.
(3) A network of large blood vessels exists in each lung.
(4) The larynx plays an important role in the exchange of oxygen and carbon dioxide.
(5) Oxygen molecules are smaller than carbon dioxide molecules.

2. Medical researchers have observed that people who smoke do not absorb oxygen from the air as effectively as people who do not smoke. Which hypothesis could be a reasonable explanation for this observation?

(1) Compounds in the smoke interfere with the functioning of the pharynx.
(2) Smoking decreases the effectiveness of carbon dioxide in the body.
(3) The smoke damages the alveoli.
(4) People who smoke tend to also get less exercise, so their hearts do not beat as fast.
(5) Smokers have more alveoli than nonsmokers.

3. What would be the predicted effect of an accumulation of liquid in the bronchioles?

(1) reduced transfer of oxygen to the capillaries
(2) more efficient removal of carbon dioxide
(3) irritation in the nose and pharynx
(4) slower heart rate
(5) growth of more alveoli

Questions 4 through 6 refer to the following passage and diagram.

Migration is the regular, often yearly, movement of animals from one place to another. Biologists have hypothesized that migrations are somehow related to the animals' need for food. It has been observed animals often commute between cooler breeding areas and tropical winter quarters near the equator. This map shows the data collected about long-distance migratory routes of several birds, one insect, and two mammals.

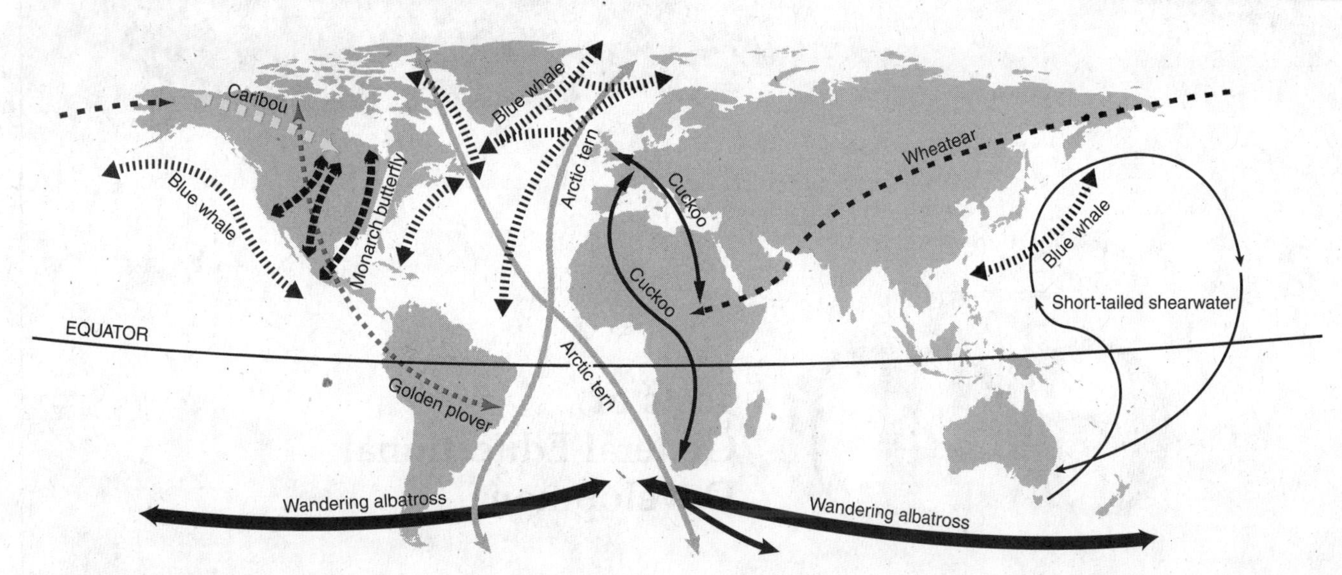

4. How could scientists determine that there are two cuckoo migrations and not just one complex migration pattern?

(1) Follow each bird to find where it travels.
(2) Use tags to identify birds from several places and find out where they migrate.
(3) Measure the lengths of birds at various places and compare them.
(4) Count the number of birds at each location.
(5) Analyze the feeding habits of different birds.

5. Which is the most likely reason that many migratory animals return to the tropics during the winter?

(1) lack of food in cold winter areas
(2) warm winter weather in the tropics
(3) desire to breed in a warm area
(4) increased competition for food in the tropics
(5) increase in natural enemies in cold winter areas

6. Which of the following statements is the best prediction based on the data in the passage and on the map?

(1) Arctic terns will migrate southward in March and northward in September.
(2) Most monarch butterflies will stay in one location their entire lives.
(3) Monarch butterflies will fly south in March and April.
(4) Short-tailed shearwaters will begin their migration in March or April.
(5) There is a single population of blue whales on Earth that moves throughout the oceans.

TIP

When you use the scientific method, remember that its overall purpose is to find ways to collect data that will explain an observation and the predictions based on that observation.

Answers and explanations start on page 115.

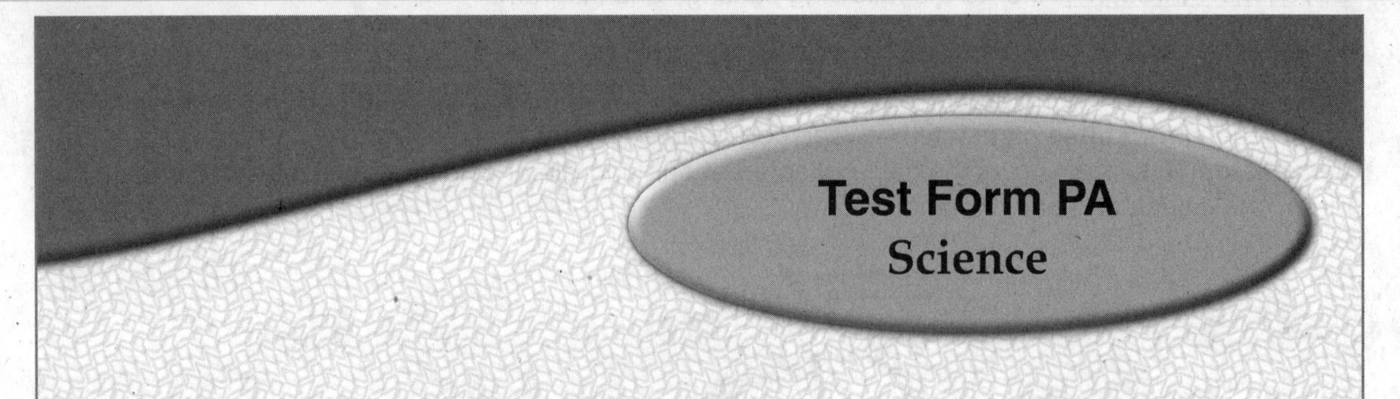

Tests of
General Educational
Development

Science
Official GED Practice Test

GED Testing Service
American Council on Education

SCIENCE

Tests of General Educational Development

Directions

The Science Test consists of multiple-choice questions intended to measure general concepts in science. The questions are based on short readings that often include a graph, chart, or figure. Study the information given and then answer the question(s) following it. Refer to the information as often as necessary in answering the questions.

You will have 40 minutes to answer the 25 questions in this booklet. Work carefully, but do not spend too much time on any one question. Be sure you answer every question.

Do not mark in this test booklet. Record your answers on the separate answer sheet provided. Be sure that all requested information is properly recorded on the answer sheet.

To record your answers, fill in the numbered circle on the answer sheet that corresponds to the answer you select for each question in the test booklet.

FOR EXAMPLE:

Which of the following is the smallest unit in a living thing?

(1) tissue
(2) organ
(3) cell
(4) muscle
(5) capillary

The correct answer is "cell"; therefore, answer space 3 would be marked on the answer sheet.

Do not rest the point of your pencil on the answer sheet while you are considering your answer. Make no stray or unnecessary marks. If you change an answer, erase your first mark completely. Mark only <u>one</u> answer space for each question; multiple answers will be scored as incorrect. Do not fold or crease your answer sheet. Return all test materials to the test administrator.

DO NOT BEGIN TAKING THIS TEST UNTIL TOLD TO DO SO

Component: 9993949132
Kit: **ISBN 0-7398-5433-X**

Directions: Choose the one best answer to each question.

1. Based on the data provided in the chart, in which months is the rainfall about the same?

Rainfall in the Rain Forest

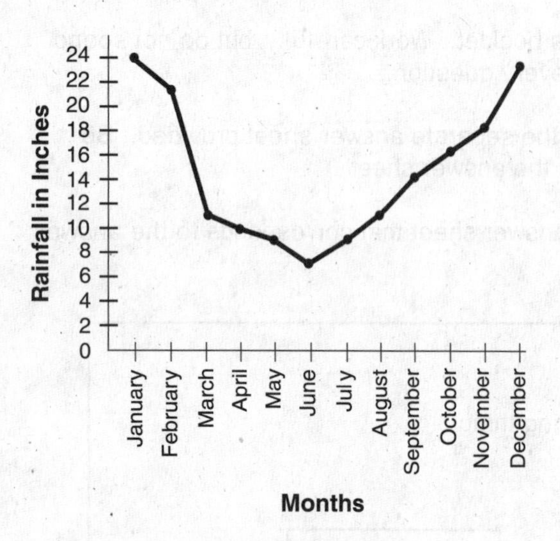

(1) March and August
(2) January and November
(3) February and June
(4) May and September
(5) October and December

2. A naturalist wanted to answer the question, "How many petals does an average daisy have?" He gathered a large number of daisies and counted the number of petals on each. He then listed his observations in the following graph.

A Survey of the Number of Petals Produced by Daisy Flowers

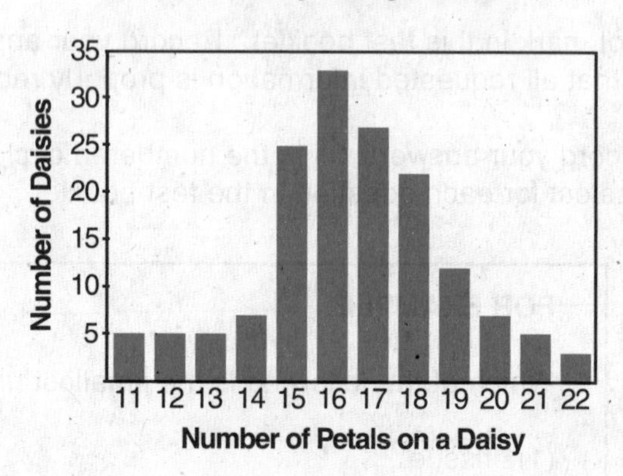

Which of the following statements could **BEST** be supported by his observations?

(1) The most common number of petals on a daisy is 15.
(2) The least common number of petals on a daisy is 20.
(3) The largest number of petals on a daisy is 19.
(4) The smallest number of petals on a daisy is 14.
(5) The number of petals on a daisy varies.

GO ON TO THE NEXT PAGE

3. Which of the following statements could be directly derived from the fact that Earth rotates on a tilted axis while revolving about the Sun?

   (1) Earth is widest at the equator.
   (2) While the Northern Hemisphere experiences winter, the Southern Hemisphere experiences summer.
   (3) Most of Earth's surface is covered by ocean.
   (4) The desert area of East Africa increases in size every year.
   (5) Erosion occurs in a west-to-east pattern.

4. At high altitudes, the atmosphere contains fewer molecules per unit volume of air than it does at low altitudes.

   For which reason may people experience shortness of breath more quickly at the top of a mountain than along a seashore?

   (1) a slower pulse rate
   (2) a greater gravitational force on the body
   (3) a lower percent of oxygen in the blood
   (4) a faster heartbeat
   (5) a slower circulation of blood

5. As a moist air mass begins to ascend one side of a mountain, the cooler, high altitudes cause the water vapor to condense and fall onto the mountain in the form of rain, hail, or snow. After the condensation occurs, the now-dry air mass continues on across the mountain.

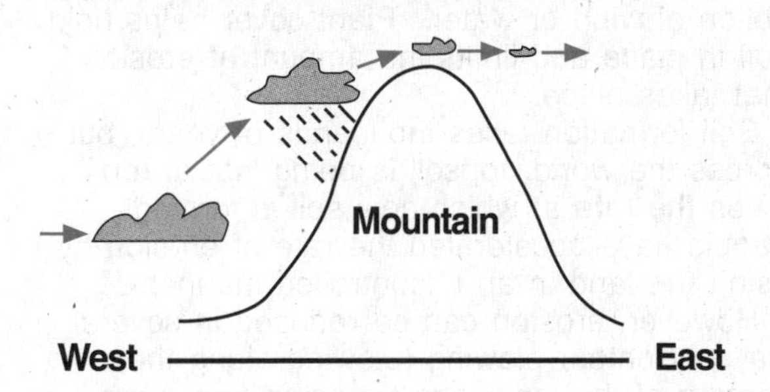

   A certain mountain range runs from north to south across a continent. At this location, the winds always blow from the west to the east. Based on the process described above, which represents the **BEST** description of the location involved?

   (1) much vegetation on the west side of the mountain range and dry conditions on the east side
   (2) desert on both sides of the mountain range
   (3) a large lake on the east side of the mountain range
   (4) tropical conditions in the northern part of the continent
   (5) desert conditions on the west side of the mountain range and lush forests on the east side

GO ON TO THE NEXT PAGE

6 **Science**

Questions 6 through 9 refer to the following information.

Agriculture depends on a layer of soil that averages only 15 centimeters in depth over Earth's surface. Crop plants rely on this rich upper layer called "topsoil." Erosion is a natural process by which topsoil is removed by the action of wind or water. Plant cover helps hold soil in place and limits the amount of erosion that takes place.

Soil formation takes thousands of years, but across the world, topsoil is being lost at ten times the rate at which new soil is formed. People have accelerated the rate of erosion by using the land in an uncontrolled manner.

However, erosion can be reduced in several ways. Contour plowing (plowing along the contour of the land) and terracing (making a series of level plots in a steplike fashion along a slope) reduce water runoff. To minimize soil loss when crop plants are spaced far apart, a method called strip cropping is used. In strip cropping, farmers grow low strips of vegetation that hold down the soil between the crops. Trees called windbreaks are planted between fields to help prevent the wind from carrying away topsoil.

Soil depletion also threatens topsoil. In a natural setting, nutrients from plants are returned to the soil as a result of decay. Farmers use fertilizers to return nutrients to the soil. A technique of alternating crops, called crop rotation, can also return nutrients to the soil. Legumes such as alfalfa and beans can add nitrogen to the soil. Legumes can be alternately grown with plants like wheat or sorghum, which deplete nitrogen from the soil.

6. H.H. Bennett, one-time chief of the Soil Conservation Service, is known as the father of soil conservation. He once said, "Productive soil is life, and the production of soil is vanishing with each passing year."

   Which solution would be the **MOST** beneficial to help resolve this environmental problem?

   (1) Tax farmers for every acre of the land they use.
   (2) Convince farmers to use accepted methods to prevent erosion.
   (3) Do not let farmers use the land.
   (4) Encourage landowners to clear-cut all vegetation.
   (5) Lower the price of fertilizers to reduce the total cost to the farmer.

7. If corn is grown in the same soil for many years, such corn will deplete nitrogen from the soil. Which method could help to return depleted nutrients to the soil?

   (1) contour plowing
   (2) terracing
   (3) crop rotation
   (4) strip cropping
   (5) windbreaks

8. Which method is used to counteract the effects of erosion caused by water?

   (1) crop rotation
   (2) soil depletion
   (3) fertilizers
   (4) legumes
   (5) terracing

9. Why are plants able to slow down soil erosion?

   (1) They prevent plant disease.
   (2) They prevent water from evaporating.
   (3) They use up nutrients in the soil.
   (4) Their roots hold the soil.
   (5) Their leaves catch water.

GO ON TO THE NEXT PAGE

10. The living cells that conduct food and water up and down through the trunk of a tree are located in a relatively thin layer just under the bark. The center part of the trunk contains old, dead cells.

    If it were necessary to kill a mature tree, which of the following methods would be most effective?

    (1) Saw a deep cut about halfway into the trunk on one side.
    (2) Remove most of the center of the trunk through a small opening.
    (3) Drive a steel spike through the bark to the center of the trunk.
    (4) Bore a hole through the center of the trunk.
    (5) Cut a deep groove below the bark around the tree trunk.

11. The blocks in the figure below each float in water and are all the same size; each is constructed of a different substance. A block will float only if its weight is less than the weight of an equal volume of water that it displaces.

    Buoyancy

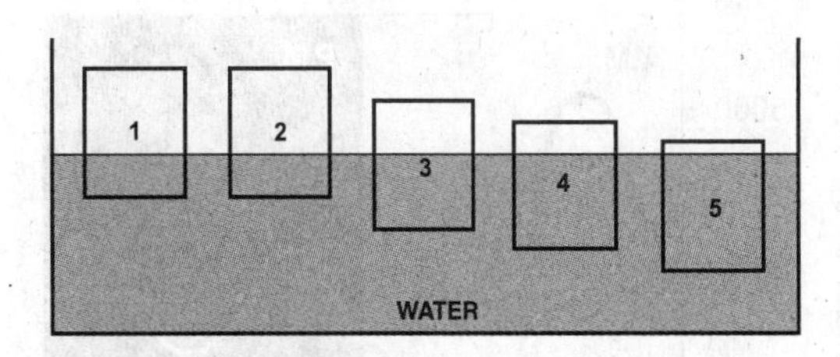

    Which of the blocks shown above displaces the largest volume of water?

    (1) Block 1
    (2) Block 2
    (3) Block 3
    (4) Block 4
    (5) Block 5

GO ON TO THE NEXT PAGE

12. Day-night rhythms dramatically affect our bodies. Probably no body system is more influenced than the nervous system. This figure illustrates the number of errors made by shift workers in different portions of the 24-hour cycle.

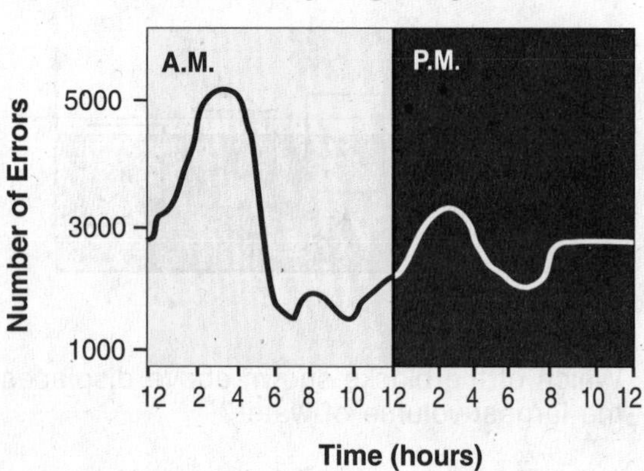

Based on the data illustrated in the figure, during which of these time periods did the most errors occur?

(1) 2 A.M. to 4 A.M.
(2) 8 A.M. to 10 A.M.
(3) 12 P.M. to 2 P.M.
(4) 2 P.M. to 4 P.M.
(5) 8 P.M. to 10 P.M.

13. As part of a laboratory experiment, five students measured the weight of the same leaf four times. They recorded 20 slightly different weights. All of the work was done carefully and correctly. Their goal was to be as accurate as possible and reduce error in the experiment to a minimum.

Which of the following is the **BEST** method to report the weight of the leaf?

(1) Ask the teacher to weigh the leaf.
(2) Report the first measurement.
(3) Average all of the weights that were recorded.
(4) Average the highest and lowest weights recorded.
(5) Discard the lowest five weights.

14. Smokestacks used by industries that burn coal or oil often give off sulfur dioxide as a by-product. Sulfur dioxide reacts with the oxygen in the air to form sulfur trioxide. Sulfur trioxide then combines with the water in the air to form sulfuric acid.

What is the most prominent atmospheric consequence of this series of chemical reactions?

(1) The atmosphere is polluted with substances that are harmful to humans and to the environment.
(2) Plants that require a basic soil with a high pH level thrive.
(3) The atmosphere becomes less polluted because sulfuric acid dissolves all particles in the air.
(4) Marine life reproduces faster because of the increasing acidity of the water.
(5) Increased sulfur dioxide emissions have little effect because wind blows the emissions away.

GO ON TO THE NEXT PAGE

Question 15 refers to the following table.

Density of Some Gases at Standard Temperature and Pressure

Gas	Density (grams per liter)
air (dry)	1.2929
ammonia	0.771
chlorine	3.214
helium	0.1785
oxygen	1.429

15. Which gas listed above is the **LEAST** dense?

(1) air (dry)
(2) ammonia
(3) chlorine
(4) helium
(5) oxygen

16. Bats navigate by sending out and receiving sound waves. The bat can determine the shape, size, and location of an object by measuring the time it takes for the sound waves to return.

Which of the following uses this principle?

(1) a telephone call that is transmitted via satellite
(2) a radio signal that is sent from a tower to a radio receiver
(3) a special photograph taken by a satellite that shows areas of different temperature
(4) a message transmitted over phone lines from one computer to another
(5) a sonar system that can determine if there are fish beneath a ship

17. The graph illustrates the cooling rates of two solutions of 150 ml of potassium nitrate (KNO_3), using a 250-ml glass beaker with a cardboard cover and an uncovered 250-ml glass beaker.

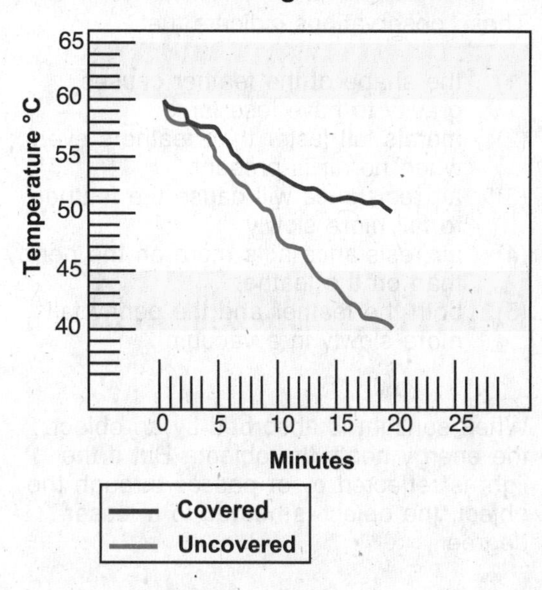

What is the **BEST** conclusion about cooling rates of solutions based on the data available in the graph?

(1) The covered solution cools more rapidly.
(2) The uncovered solution cools more rapidly.
(3) The concentration of the solution affects the cooling rate.
(4) The original temperature of the solution affects the cooling rate.
(5) The volume of the beakers affects the cooling rate.

GO ON TO THE NEXT PAGE

18. The force of gravity causes objects to fall toward Earth, but it is commonly observed that a feather falls more slowly than a penny. However, when a feather and a penny are placed in a chamber where all the air has been removed, they both fall at the same rate.

    These observations indicate that

    (1) the shape of the feather causes gravity to have less force
    (2) metals fall faster than feathers even when no air is present
    (3) air resistance will cause the feather to fall more slowly
    (4) air resistance acts more on the penny than on the feather
    (5) both the feather and the penny fall more slowly in a vacuum

19. When sunlight is absorbed by an object, the energy heats the object. But if the light is reflected by or passes through the object, the object is heated to a lesser degree.

    An automobile with black seatcovers is left outside on a sunny day with its windows rolled up. Which of the following will heat up the **MOST** while causing the inside of the automobile to get warm?

    (1) window glass in the side windows
    (2) the white steering wheel cover
    (3) window glass in the front windshield
    (4) the black seatcovers
    (5) the air inside the automobile

20. The graph below shows the percentage of light of different wavelengths that is absorbed by chlorophyll.

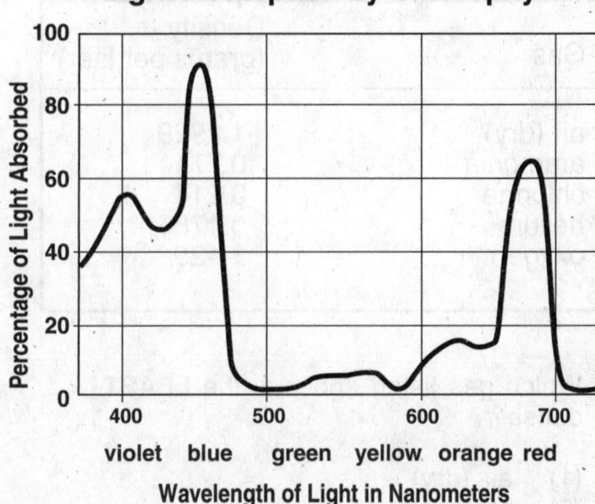

    For a plant to achieve the maximum rate of photosynthesis, what wavelengths of light would be **MOST** effective?

    (1) violet and blue
    (2) blue and red
    (3) green and orange
    (4) orange and red
    (5) violet and yellow

GO ON TO THE NEXT PAGE

21. As many as 20 percent of patients who take prescription drugs including insulin also consume herbal supplements, but without realizing that such a combination may pose a health risk under certain circumstances. Patients should always consult their pharmacists or physicians before taking herbal supplements with prescription drugs.

Herb Interactions

HERB	INTERACTION	WHEN TAKEN WITH
Garlic	Increased internal bleeding	Blood-thinning drugs
Gingko Biloba	Increased internal bleeding	Blood-thinning drugs
St. John's Wort	Increased sedation	Prozac
Ginseng	Lowered blood-sugar level	Insulin or blood-sugar regulators

From information provided in the chart above, which of the following statements is true?

Herbal supplements

(1) are types of vitamins that have a high calorie content
(2) interfere with all prescription drugs
(3) are cheaper than prescription drugs
(4) boost the human immune system
(5) may change the effect of certain prescription drugs

22. Two identical balls are suspended by a rod, as shown in the diagram below.

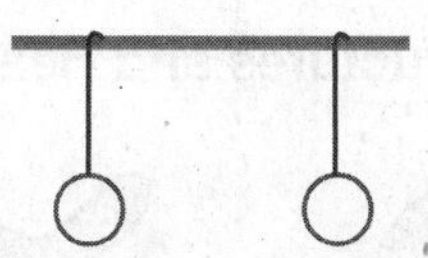

A positive electrical charge is put on the left ball, and a negative charge on the ball on the right.

Expected Change in Position When Particles Have Opposite Charges

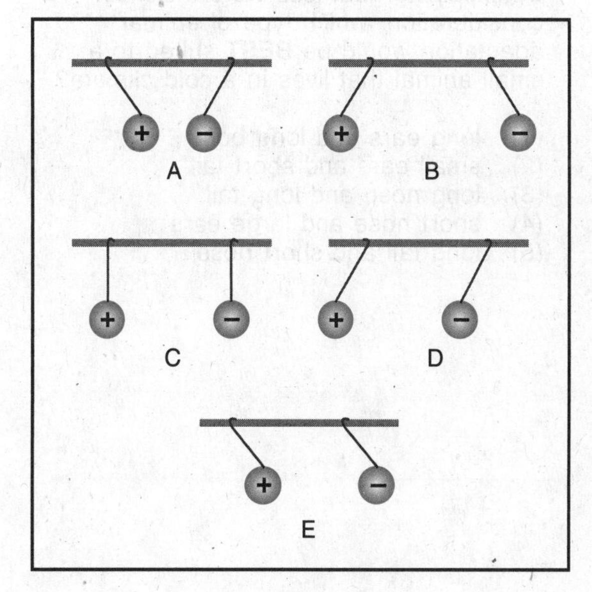

After placing the charges on the balls, which of the following **BEST** represents how the balls will react to each other?

(1) A
(2) B
(3) C
(4) D
(5) E

GO ON TO THE NEXT PAGE

23. If object A and object B below both have the same mass, object B will lose heat more quickly than object A.

Body Structures and Heat Loss

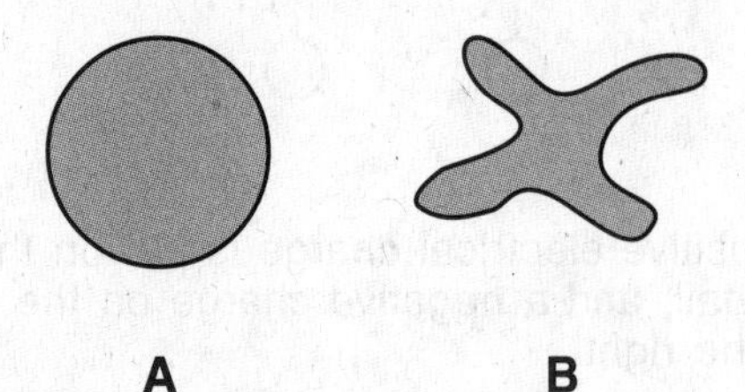

If minimizing heat loss were the main consideration, which type of animal adaptation would be **BEST** suited to a small animal that lives in a cold climate?

(1) long ears and long body
(2) small ears and short tail
(3) long nose and long tail
(4) short nose and large ears
(5) long tail and short nose

24. In order to cut her grass, Georgette recently purchased a string trimmer with the following instructions.

Mixing Instructions for 2-Cycle Engine Oil

Mix 2-cycle oil with unleaded gasoline in a 24:1 ratio of gasoline (gallons) to oil (ounces). Use the mixing instructions from an 8-ounce container of 2-cycle engine oil as listed in the following table.

Ratio	16:1	20:1	24:1	32:1	40:1
Gas (Gal)	1.0	1.25	1.5	2.0	2.5
Oil (Oz)	8.0	8.0	8.0	8.0	8.0

Georgette needs to fill the trimmer's engine properly before she uses it.

Which amount of gasoline should she mix with the contents of the oil container?

(1) 1.0 gallon
(2) 1.25 gallons
(3) 1.5 gallons
(4) 2.0 gallons
(5) 2.5 gallons

GO ON TO THE NEXT PAGE

25. Earth's environment has changed over the centuries. Such changes can be viewed by examining fossils below Earth's surface. The fossils shown in the following diagram were found in a rock cliff.

Location of Fossils in Earth's Layers

	Layer	Fossils
Surface	D	leaf, plant stem, animal footprints
↓	C	tree trunk, flower stem, animal footprints
	B	shark teeth, fish scales, plant stem
Deep Below Surface	A	plant stem, leaf, flower prints

What environmental change **BEST** explains the differences between layers A and B?

(1) The weather changed.
(2) Oceans covered the area.
(3) Earthquakes occurred.
(4) No observable changes are present.
(5) Drought conditions prevailed.

END OF EXAMINATION

To determine the standard score for the *Official GED Practice Test Form PA: Science:*

1. Locate the number of questions the candidate answered correctly on the multiple-choice test.
2. Read the corresponding standard score from the column on the right.

Compare the candidate's standard scores to the minimum score requirements in the jurisdiction in which the GED credential is to be issued. (See *Appendix D* in the *Official GED Practice Tests Administrator's Manual.*)

U.S. Edition Form PA Science	
Number of Correct Answers	Estimated GED Test Standard Score
25	800
24	620
23	550
22	520
21	490
20	470
19	450
18	440
17	430
16	420
15	420
14	410
13	400
12	390
11	380
10	370
9	360
8	350
7	330
6	320
5	300
4	290
3	260
2	240
1	200

Science Answers

1. 1
2. 5
3. 2
4. 3
5. 1
6. 2
7. 3
8. 5
9. 4
10. 5
11. 5
12. 1
13. 3
14. 1
15. 4
16. 5
17. 2
18. 3
19. 4
20. 2
21. 5
22. 1
23. 2
24. 3
25. 2

Pretest Answers and Explanations

1. **(2) tennis ball rolling on a grassy lawn** As the law of physics states, the resistance to motion is the greatest when there is a force acting against the motion. Option (2) is correct because the grass of the lawn is acting as a force against the motion of the tennis ball. The grass is the greatest force in all the options. The other options are not correct because there is no force working against the motion of the bowling ball (option 1) or the car (option 5). In option (3), the wind is actually increasing the motion of the leaves, not reducing motion. The motion of the rock falling off the cliff (option 4) is also increased by the pull of gravity.
2. **(3) Light energy travels faster than sound energy.** Because the visual information arrives first, you can infer that light travels faster than sound. Light and sound traveling at the same speed (option 1) contradicts the described observation. The suggestions that sound occurs after the flash (option 2) and that ears process information several seconds faster than eyes (option 5) contradict everyday experiences and observations. Sound energy following a longer wave path (option 4) could explain a time difference, but there is no indication in the passage or normal experience to indicate that this happens.
3. **(4) Ungulates (hoofed mammals) have four legs.** The bar on the graph corresponding to ungulates indicates that they have four legs, so option (4) is correct. The heading states that the graph applies to selected animals, not all animals (option 1). The graph does not present any information about the speeds of animals (option 2) or the complexity of their bodies (option 3). The length of the bar shows that crabs have 10 legs, not 8 (option 5).
4. **(2) Test the prediction using other nitrate and bromide salts.** Because the hypothesis is that solubility depends on the negative ion, the prediction must be tested with salts having the same metal ions, but different nitrate and bromide ions, so option (2) is correct. The student does not have enough data to draw a conclusion (option 1) or report final results (option 3). Repeating the experiment with the same salts does not test the prediction (option 4). Performing the experiment with different temperatures (option 5) does not provide useful data to test the student's hypothesis—the salt is the variable, not temperature.
5. **(3) 46°C** The correct answer is option (3) because the two lines representing solubility cross at that temperature. When lines on a line graph cross, they have the same value at that point. At temperatures less than 46°C (options 1 and 2), KBr is more soluble than KNO_3. At temperatures greater than 46°C (options 4 and 5), KNO_3 has a greater solubility than KBr.
6. **(2) An accepted, orderly system aids scientific learning and communication.** Although the passage does not directly state that the system aids learning and communication, it provides information that leads to that conclusion. Linnaeus's system allowed scientists to distinguish animals based on scientific names, so option (2) is correct. The passage states that there was research on organisms before 1735 (option 1). There is no indication that all scientific communication is in Latin (option 3) or that the scientific name for humans is *Homo sapiens* (option 4). Option (5) contradicts the passage.
7. **(3) 1.52** Option 3 is correct because in the distance column, the table shows that Mars is 1.52 AU from the sun. The other options are incorrect readings of the graph. Option (1) is the distance in AU from Mercury to the sun. Option (2) is the distance in AU from Earth to the sun. Option (4) is the distance from Jupiter to the sun in AU. Option (5) is the rotation of Mars in hours.
8. **(2) The period of a planet's revolution increases as its distance from the sun increases.** Evaluation of the information in the table shows that both distance and period of revolution increase from top to bottom, so option (2) is correct. The periods of rotation do not have any apparent correlation to distance (option 1) or periods of revolution (option 5). While option (3) is a true statement, it is not based on information in the table because the sizes of the planets are not given. Option (4) contradicts the information in the table.
9. **(5) Living things exhibit behaviors as a response to their environment.** Option (5) is correct because the passage describes responses to the environment and defines them collectively as behavior. Plants growing toward light (option 1) and people responding to cold weather (option 4) are supporting facts, not the main idea. Option (2) contradicts the first sentence of the passage, which states that all living things respond to their environment. Bacteria and plants are living things (option 3), but this is a detail and not the main idea presented in the passage.
10. **(4) The DNA of animal cells is contained in the nucleus, but virus DNA is not.** Labels on the diagrams show that the DNA of an animal cell is in its nucleus while the DNA of a virus is surrounded by a protein coat, so option (4) is correct. The diagrams do not indicate the amount of DNA (option 1) or its complexity (option 2). The statement that the DNA molecules contain different bases (option 3) is not true and does not relate to the diagrams. Option (5) contradicts the diagrams.
11. **(3) Burning fossil fuels releases much more carbon dioxide into the atmosphere than breathing does, so the effects of the activities are not the same.** Option (3) is correct because global warming is caused by an excess of carbon dioxide, and fossil fuel combustion releases excessive amounts of carbon dioxide, while breathing releases only trace amounts. The statement that people do not release carbon dioxide (option 1) contradicts the passage. There is no difference between molecules of carbon dioxide from different sources (options 2 and 4). The heat from combustion (option 5) is not mentioned in the argument.
12. **(5) about 24 times fewer vertebrates than invertebrates** Option (5) is correct because according to the graph, invertebrate species, including insects, far outnumber vertebrate species. In fact, only 4% of all animals are vertebrates and 96% are invertebrates. This means there are about 24 times fewer vertebrates than invertebrates (96% divided by 4%). The numbers given in options (1) and (2) show that there are fewer vertebrates than invertebrates, but the differences given are too low. Options (3) and (4) indicate that there are more vertebrates than invertebrates, which contradicts the information in the graph.
13. **(4) The risks of genetic modification are great enough that it should not be used.** Option (4) is correct. The passage states that some scientists *think* the

risks are too great, indicating an opinion. That genetic modification is a relatively new technique (option 1), that it changes specific traits (option 2), and that traits can be transferred (option 5) are facts stated in the passage. The statement that genetic modification has not been successful (option 3) contradicts the facts of the passage.

14. **(3) Comets normally have very distant orbits, but some are pulled out of orbit by gravity and pass close to the sun.** Option (3) is correct because the passage describes the motions of comets that approach the sun. The statements that comets are part of the solar system (option 1) and that Jupiter causes Kuiper Belt comets to move toward the sun (option 2) are facts/details from the passage but not adequate summaries of the entire passage. The composition of comets (option 4) only relates to the first paragraph. The statement that comets crash into the sun (option 5) contradicts the passage.

15. **(3) gravitational effects of a passing star** The third paragraph states that gravitational effects from passing stars are the cause of Oort Cloud comets moving toward the sun, so option (3) is correct. The sun pulls the comet (option 1) but does not cause it to move from its original orbit. The passage states that Jupiter's gravitational pull (option 2) is not strong enough to cause comets to move from the Oort Cloud. Neptune (option 5) is not mentioned in the paragraph about the Oort Cloud, and it exerts an even smaller pull than Jupiter. Evaporation (option 4) is an effect of approaching the sun, not a cause.

16. **(3) The vast majority of Earth's water is in the oceans, and most of the rest is held in icecaps, glaciers, and inland seas.** Option (3) is correct because the first bar of the diagram shows that 97% of Earth's water is in the oceans, and the second bar shows that 77% of the remaining water is in icecaps, glaciers, and inland seas. The diagram does not indicate distribution on the surface (option 1). Option (2) is incorrect because lakes and rivers account for about 61% of freshwater, but that is less than 1% of the total water on Earth. There is no information in the diagram about sources of drinking water (option 4) or evaporation (option 5).

17. **(3) Alpha radiation is positively charged, and beta radiation is negatively charged.** Option (3) is correct because opposite charges attract. Alpha radiation is deflected toward the negative plate, indicating a positive charge, and beta radiation is deflected toward the positive plate, indicating a negative charge. If both particles had the same charge (options 1 and 2), they would be deflected in the same direction. The type of charges can be determined by their interaction with the plates (option 4). There is no indication of gaining charges as particles pass the plates (option 5).

18. **(2) Opposite charges attract and like charges repel.** Option (2) is correct. The effect of charges on one another is a key point of the diagram that is not stated, so the writer assumes that the reader already knows the effect. The facts that lead stops radiation (option 1) and that gamma rays are electromagnetic radiation (option 3) are true but not necessary to understand the diagram or passage. The amount of deflection (option 4) is not relevant to understanding the concepts. That there are three types of radiation (option 5) is clearly stated and is not an assumption.

19. **(2) The acceleration due to gravity is the same for any falling object.** The information in the passage is adequate to support the statement that the acceleration due to gravity is the same for all objects (option 2), as shown by the value 9.8 m/s^2. The statements that gravity is the only force acting on a falling object (option 1) and that wind resistance explains why a feather falls slower than a marble (option 5) are not adequately supported by facts in the passage. The passage states that Galileo might not have dropped objects from the tower (option 3), so the statement is inaccurate. The force of gravity applies to all objects (option 4), so the statement is not accurate.

20. **(4) Tropical regions have the greatest occurrence of malaria.** The map shows that malaria is transmitted in regions near the equator, which are tropical regions (option 4). The map contradicts the restatement that most of the population is not at risk for malaria (option 1). Option (2) is incorrect because it contradicts the map, which shows that malaria has been eradicated in the United States and Canada. There is no data to support that Australia's status is likely to change (option 3). The passage indicates how malaria is transmitted, but the map does not indicate how malaria is transmitted (option 5), only where it is transmitted.

Answers and Explanations

Skill 1 Identify the Main Idea

Pages 12–13

1. **(2) It has four particles in the nucleus.** Option (2) is correct because the diagram shows the helium nucleus with two protons and two neutrons. There are only two neutrons, not four (option 1). There are two, not one, electrons that move around the nucleus (option 3). Two electrons move around the nucleus, not protons (option 4). The atom has the same number of protons and electrons, not more protons (option 5).
2. **(4) Elements are made up of tiny particles called atoms.** Option (4) is the right answer because the first sentence states that elements are made of atoms, and the rest of the paragraph discusses the makeup of atoms. The other choices consist of information that is not in the first paragraph (options 1 and 2) or details that only support the main idea (options 3 and 5).
3. **(3) Three of the most important particles in atoms are electrons, protons, and neutrons.** Option (3) is the correct answer because it is stated in the first sentence of the paragraph and the rest of the paragraph describes these three particles. Option (1) is wrong because it contains information that is in the first paragraph, not the second. The other choices are wrong because they are supporting details and not the main idea of the second paragraph (options 2, 4, and 5).
4. **(2) PolyAspirin works like aspirin but is safer for the stomach.** The main idea of the passage—PolyAspirin is a safer drug that works like aspirin—is presented in the first sentence of the second paragraph (option 2). The other choices are all details or examples that only support the main idea (options 1, 3, 4, and 5).
5. **(1) PolyAspirin is a chain that breaks down into smaller molecules, including salicylic acid.** The main idea of the diagram, that PolyAspirin breaks down into smaller molecules (option 1), is shown by the structure of the chain and the key, which shows salicylic acid is formed by the breakdown. The structure of the molecules in the chain (options 2 and 5) is supporting information for the main idea. Option (3) presents information from the passage that is not included in the diagram. Option (4) is an opinion based on the passage.
6. **(5) Glaciers can cause massive changes in the landscape.** Option (5) is correct. The main idea of the passage—that glaciers are powerful forces that shape Earth's surface—is stated in the first sentence. The fact that they are made of ice and snow (option 1) and that they carved cliffs in Yosemite Park (option 3) is supporting information. Most glaciers move less than 8 miles per year (option 2), according to the passage. The size of glaciers (option 4) was not mentioned.

Skill 2 Restate Information Pages 16–17

1. **(4) large waste particles** Option (4) is correct because according to the passage, wastes move out of the cell, which is the process of exocytosis, as shown in the diagram. Food (option 1) and water (option 2) move into the cell during endocytosis. There is no indication that the nucleus moves (option 3), and oxygen is not mentioned in the passage or the illustration (option 5).
2. **(1) The cell membrane surrounds large particles.** The diagram shows that the cell membrane surrounds particles during endocytosis; the passage indicates that they are large, so option (1) correctly restates this process. Absorption of small particles (option 2) contradicts the passage. Expelling particles (options 3 and 4) contradicts the endocytosis diagram. Endocytosis is a method of allowing, not preventing (option 5), particles to enter the cell.
3. **(4) Acceleration can be calculated by dividing the force by the mass.** Option (4) is the correct answer because it correctly restates the equation from the diagram in words. The ball does accelerate (option 1), but that does not show the relationship of the three values. There is no information about ease of hitting the ball (option 2). Option (3) contradicts the information in the diagram, which shows acceleration as the quotient, not the product. The diagram indicates that force and acceleration are in the same direction (option 5), but this choice does not address the relationship with mass.
4. **(5) Greenhouse gases trap energy from the sun within the earth's atmosphere.** Option (5) correctly restates information from the first paragraph and the diagram. Option (1) is incorrect because the passage states that greenhouse gases—in the right amounts—are actually beneficial to Earth. Option (2) is a supporting detail and is too specific to be a restatement. Option (3) is true, but this information is found only in the passage, not in the diagram. Option (4) is the main idea of the second paragraph, not the first.
5. **(2) Global warming can cause changes in the weather.** Option (2) is correct because it restates the sentence in the passage about how global warming can cause changes in climate. The passage says that the greenhouse effect, not global warming (option 1), supports life. Reflection of energy and light (options 3 and 4) are not related to global warming, according to the passage. Reduction of oxygen (option 5) is not mentioned.
6. **(2) Each square of the periodic table represents an element, and the squares are organized based on structures and properties of the elements.** Option (2) concisely restates the information from the passage and diagram, only mentioning the most important information. The number of periods and groups (option 1) and number of elements (option 5) are correct information, but they only pertain to a specific part of the given information. It is incorrect that the table includes all information about the elements (option 3). The structure of atoms (option 4) is too specific to be a restatement.

Skill 3 Summarize Ideas Pages 20–21

1. **(5) Carbon dioxide levels and global temperatures generally increased.** The general trend of both lines on the graph is upward, which indicates an increase (option 5). While carbon dioxide levels did increase (option 1), this is not a complete summary because it does not address global temperatures. Carbon dioxide levels have only increased, not decreased (option 2). Global temperatures have fluctuated only a small amount (option 3), but they have been generally increasing. The upward trend shown in the graph means that global temperatures have not decreased (option 4).

2. **(5) Carbon dioxide levels and global temperatures will continue to rise.** The passage indicates that carbon dioxide levels will rise, and the diagram indicates that as carbon dioxide increases, so will global temperatures (option 5). A decrease in carbon dioxide (option 1), a drop in global temperatures (option 3), and an end to global warming (option 4) contradict the passage and diagram. There is no information to indicate that gases will stop trapping the sun's heat (option 2).
3. **(4) The four planets have temperatures ranging from –23°C to 460°C.** The table shows data for four planets, so a summary should describe the data about all of them (option 4). The other answer choices (options 1, 2, 3, and 5) are all correct, but each includes only one detail from the table.
4. **(2) DNA molecules store instructions for cell growth and genetic information in a pattern of molecules.** This statement (option 2) is a correct, brief summary of the key ideas. The shape of the DNA molecule (option 1) and its ability to divide (option 4) are details and do not provide enough information to summarize the passage. The other statements (options 3 and 5) are not relevant to the information in the passage.
5. **(4) A DNA molecule can divide into two parts by separating along the central linkages.** The key idea of the illustration is the separation of the two helices (option 4). DNA's role in cell reproduction (option 1) and the ability to store genetic information (option 5) are related to the passage, not the illustration. No information is given about plants or animals (option 2). The length of the molecule (option 3) is not the key point and is actually drawn much larger than its actual size.
6. **(3) Only banning CFCs will reduce the amount of chlorine in the atmosphere.** This statement (option 3) correctly and briefly expresses the key point of the graph in words. The graph does not include information about the ozone layer (option 1). Options (2), (4), and (5) are incorrect because they contradict the graph, which shows that only banning CFCs will reduce chlorine levels.

Skill 4 Identify Implications
Pages 24–25

1. **(4) It originates inside the plant.** The source of all the other ingredients comes from outside the leaf, as shown by the purple arrows. This implies that chlorophyll is already in the leaf (option 4). The diagram states that water, not chlorophyll (option 1), is broken down into hydrogen and oxygen. The diagram shows that oxygen leaves the leaf, so option (2) is not correct. There is no indication that anything is made from carbon and oxygen (option 3) or that chlorophyll leaves the plant (option 5) during photosynthesis.
2. **(1) water and oxygen** The only two arrows leaving the leaf are the blue arrows for oxygen and water (option 1), which implies that they are the only two results released during photosynthesis. Hydrogen, chlorophyll, carbon, and glucose (options 2, 3, 4, and 5) are not released as byproducts.
3. **(1) Exposure to very large amounts of microwaves should be avoided.** The passage implies that large amounts of microwaves should be avoided because they can cause harm (option 1). The passage states X-rays can cause radiation burns, but skin cancer is not mentioned, so option (4) is not correct. The statements that humans—a type of animal—are not at risk (option 2) and that microwaves from microwave ovens are not dangerous (option 3) contradict the passage. Nervous system problems can occur for a variety of reasons; avoiding exposure to microwaves would only reduce these problems, not prevent them (option 5).
4. **(5) The life cycle of a medium-sized star is fairly predictable.** Life cycles are observed patterns, which makes them predictable, and the title implies that the diagram refers to only medium-sized stars (option 5). There is no implication that it applies to all stars (option 1) or that some main sequence stars do not become giant stars (option 2). Option (3) is incorrect because Stage 4 of the diagram states that a giant star is 10 times bigger than our sun; therefore, our sun is not a giant star. Fusion ends at stage 5, so it does not occur during all stages of a star's life (option 4).
5. **(2) A star reaches its largest size during Stage 4.** The diagram shows relative sizes, with Stage 4 as the largest (option 2). There is no information about length of the stages (options 1 and 4). There is no implication that medium-sized stars begin as anything other than a nebula (option 3). The figure shows that a white dwarf turns into a black dwarf, not a planetary nebula (option 5).
6. **(3) fusion** According to the text of the diagram, fusion (option 3) begins at Stage 2 and ends at Stage 5, implying that there is no fusion during the other stages. Both expansion (option 1) and contraction (option 2) occur only during stages 2 and 4. There is no indication that the star explodes (option 4), and the figure implies that cooling (option 5) occurs only in Stage 7.
7. **(3) Our sun will increase in size by at least ten times.** The diagram indicates that main sequence stars will become giant stars and will expand to at least 10 times the size of our sun; therefore, it is implied that our sun will become a giant star and increase in size by 10 times (option 3). The sun is only at Stage 3, so its life cycle is not almost over (option 1). There is no information about the length of time between stages (options 2 and 5). The detail of the diagram implies that the cycle is well understood (option 4).

KEY Skill 5 Apply Scientific Principles
Pages 28–29

1. **(4) pruning shears, which have a fulcrum between the effort and work points** According to the figure, a class 1 lever has its fulcrum between the effort point and its work point, which is true for pruning shears (option 4). Tweezers (option 1) are class 3 levers. A paper cutter (option 2) and a nutcracker (option 5) are examples of class 2 levers. A hammer (option 3) is not a lever because it has no fulcrum.
2. **(3) Classes of levers vary in the relative positions of their fulcrums, work points, and effort points.** The figure identifies how each different lever works (option 3) and gives a real-world example for each class. The figure does not indicate relative amounts of work (options 1 and 5) performed by each lever. All of the levers in the diagram have a fulcrum (option 2), although its location varies in each class.

While a wheelbarrow has two handles, the diagram does not indicate it has two effort points (option 4).

3. **(5) Male warblers that sing more songs tend to find mates sooner than male warblers that sing fewer songs.** The downward line on the graph shows that the more songs a male warbler sings, the fewer number of days it takes for the warbler to pair with a female (option 5). Warblers do not all sing the same songs (option 1), as is evidenced by the *x*-axis title, "Number of different songs sung by male." There is no information about whether female warblers sing (option 2), or whether warblers sing at times other than mating season (option 3). The graph shows that some male warblers sing only about 21 songs and still pair with a mate; therefore, option (4) is incorrect.
4. **(2) Earthquakes can occur both on land and below the oceans.** As shown on the map, dots representing earthquakes appear on the continents and in the oceans, meaning, as a general principle, earthquakes can occur on land and below the oceans (option 2). The map does not indicate the number of earthquakes per year (option 1) or total number of earthquakes in a region (option 3). The map indicates that earthquakes do occur in the Pacific Ocean and in the central United States; therefore, options (4) and (5) are incorrect.
5. **(5) the southeast coast of South America** Option (5) is correct. If you apply the scientific principle behind tsunami creation, you can conclude that coastlines that face active earthquake zones are in the most danger of being struck by a tsunami. The southeast coast of South America does not face an active earthquake zone. Options (1), (2), (3), and (4) all face active earthquake zones and are thus in danger of being hit by a tsunami.
6. **(4) A boundary between plates exists beneath where belts or strings of earthquakes are shown on the map.** Since movements of large plates cause earthquakes, it is a scientific principle that there is a plate boundary everywhere a major belt of earthquakes exists (option 4). Option (1) is incorrect because there are no earthquake zones along the southeast coast of South America. Option (2) is incorrect; while it is true the Ring of Fire is one of Earth's major earthquake zones, this does not apply the scientific principle of earthquakes being caused by the movement of plates. Options (3) and (5) are incorrect and contradicted by the map.

KEY Skill 6 Make Inferences

Pages 32–33

1. **(4) Sponge cells store the materials that palisade cells use during photosynthesis.** Option (4) is an inference based on information that chlorophyll is contained in palisade cells and the materials for photosynthesis are stored in sponge cells. Option (1) is incorrect because photosynthesis only occurs in chlorophyll-containing cells. Although the two types of cells are in different layers, the diagram shows they are side by side and therefore do interact (option 2). Options (3) and (5) are incorrect because the passage does not indicate that either sponge or palisade cells help protect anything.
2. **(4) skin** The passage and diagram indicate the main functions of the cuticle are to protect the cells beneath it and to retain water. In humans, skin (option 4) performs this function. The other human organs (options 1, 2, 3, and 5) have completely different functions.
3. **(1) Molecules in liquid water are in motion.** The passage states that added energy makes the molecules move faster, so you can infer that they are already moving (option 1). The temperature at which water evaporates is not mentioned (option 2). The definition of water vapor is water in its gas state, not liquid (option 3). Option (4) contradicts the passage. There is nothing in the passage about the role of water for living things (option 5).
4. **(3) Different constellations are visible during different seasons.** The passage states that Orion can be seen in winter, from which you can infer that some constellations can only be seen in certain seasons (option 3). The brightest star in Orion is Rigel, which is not the closest to Earth (option 1). The passage and diagram do not address any facts from which you can infer the age of the stars (option 2). Option (4) contradicts information in the diagram. Motion of stars is not mentioned (option 5).
5. **(5) As seen from Earth, stars appear close together, but they are actually light years apart.** The passage indicates that constellations are patterns of stars that appear close together, but the diagram indicates that they can be very far apart (option 5). The diagram states the distances of specific stars but does not state the distance to every star (option 1). There is no information in the passage or diagram that would allow you to infer something about the brightness of the belt stars (option 2), the size of the stars in Orion (option 3), or the total number of constellations (option 4).
6. **(4) Chandra helps scientists learn more about the hot matter in our universe.** The passage states that very hot matter emits X-rays and Chandra detects X-rays, so you can infer that Chandra is used to look at hot matter (option 4). The passage does not indicate that Chandra completely replaced older telescopes (option 1) or that NASA will launch a new telescope soon (option 2). The invention of Chandra indicates the science is advancing, which means there is the possibility that telescopes will one day see farther than 10 billion light years, so option (3) is wrong. The year of launch is stated as fact (option 5) and does not need to be inferred.

KEY Skill 7 Identify Facts and Opinions

Pages 36–37

1. **(2) Compost is better for the garden than synthetic fertilizers.** Option (2) is correct because the phrase "some gardeners believe" indicates an opinion. That compost can enhance soil (option 1) and composting cannot occur without microorganisms (option 3) are facts stated directly in the passage. The production of heat (option 4) is a fact shown in the diagram and stated in the passage. The statement that carbon dioxide is necessary for composting to occur (option 5) contradicts the passage and diagram.
2. **(2) Composting generates carbon dioxide and heat.** Option (2) is correct because it is a fact stated in both the passage and the diagram. Option (1) is incorrect because although some people prefer compost made from kitchen waste, it is not a fact that it is better than compost made from lawn waste. Options (3) and (5) are opinions, as indicated by

the phrases *some gardeners believe* and *some people think* in the passage. Option (4) is not addressed in the passage or diagram.

3. **(5) Nuclear power plants are too dangerous to be located near cities.** Option (5) is correct because the phrase "Many people think..." in the passage indicates that this statement is an opinion. That nuclear plants are expensive (option 1) and energy is released when atoms split apart (option 2) are facts stated directly in the passage. Option (3) contradicts facts in the passage. The statement about fission being used in power plants (option 4) is not directly stated but can be concluded from other facts, and therefore, is not an opinion.
4. **(3) There are large undiscovered reserves of fossil fuels.** Option (3) is correct because the passage states that some scientists *think* there are more reserves, but this has not been proven as a fact. That coal reserves will last longer than other fossil fuels (option 1) is a conclusion based on the diagram. The statements that oil and gas are nonrenewable (option 2) and that coal is a fossil fuel (option 4) are facts from the passage. The availability of other energy sources (option 5) is a fact mentioned at the end of the passage.
5. **(5) It takes millions of years to replace fossil fuels.** Option (5) is correct because it is a stated fact from the passage. The other answer choices (options 1–4) are all unstated opinions; they are points of view that are not proven to be true by any of the information in the passage or diagram.
6. **(2) Cloud seeding can create problems for places that are downwind of the seeding.** Option (2) is correct because the statement in the passage includes the phrase "some scientists believe..." The statements that it is hard to monitor cloud seeding success (option 1) and silver iodide has a crystal structure similar to ice (option 3) are facts found in the passage. Option (4) is also a fact, not an opinion. Option (5) contradicts facts in the passage.

Skill 8 Recognize Assumptions
Pages 40–41

1. **(2) what magnets are** Option (2) is correct because the passage describes Earth as a magnet but does not explain what a magnet is. The reader does not need to understand how a compass works (option 1) or that some rocks are natural magnets (option 4) to understand the passage. Magnetic poles are defined in the passage (option 5), but it is not necessary to know how they formed (option 3).
2. **(2) The reader is familiar with the term "magnetic field."** The illustration refers to magnetic field lines (option 2) without defining magnetic field. It is not necessary to know how to calculate field strength (option 1) or to have experimented with magnets (option 3) in order to understand the diagram. Earth's magnetic field cannot be seen (option 4). Option (5) contradicts the illustration.
3. **(3) Magnetic force is represented by magnetic field lines.** The illustrator assumes that the reader realizes the magnetic field lines represent the magnetic force that surrounds the Earth (option 3). That the field extends into space (option 1) and that it has two poles (option 2) are explicitly shown in the illustration as facts, not assumptions. The magnetic field cannot be seen from space (option 4). The bar magnet is not an actual bar magnet, but if it were, it would run through the center of Earth, not be located on the surface (option 5).
4. **(1) Birds eat fish.** Option (1) is correct because the reader must understand how a food chain works: each level of the food chain provides food for the next — fish eat phytoplankton and birds eat fish. The diagram illustrates population, not size of organisms (option 2). The statement that all organisms use sunlight to produce food (option 3) is not related to the diagram and is not true. The diagram illustrates a decreasing population higher on the food chain, so option (4) contradicts the diagram. That top organisms are always birds or mammals (option 5) is not a true statement.
5. **(5) how to read Roman numerals** The groups are labeled with Roman numerals, so it is assumed that the reader knows what they mean (option 5). It is not necessary to know what magnesium looks like (option 1) to understand the table. The representation of periods (option 2) and identity of noble gases (option 4) are labeled clearly as facts; the reader would not need to already know this information. The difference in atomic number (option 3) can be determined using information from the table, so it is not assumed the reader already knows this.
6. **(4) The reader knows what atomic numbers represent.** The table indicates where atomic numbers are located, but does not define what they are (option 4), so it is assumed the person using the table knows what these numbers represent. The names of the Actinides (option 1) are stated on the table, so they are known facts, not assumptions. It is not necessary to know that some elements are radioactive (option 2) or to have studied noble gases (option 3) to understand the table, so these were not assumptions made by the creator. Some elements are gas or liquid at room temperature (option 5), so that statement is an incorrect assumption.

KEY Skill 9 Identify Causes and Effects
Pages 44–45

1. **(1) the reaction of methane and oxygen** The passage and table both indicate that methane and oxygen (option 1) react to form carbon dioxide and water. There is no information about the mixing of water and carbon (option 2), the reaction of carbon monoxide and oxygen (option 4), or changes involving nitrogen (option 5). The evaporation of natural gas (option 3) is not a chemical reaction.
2. **(2) formation of CO_2** Option (2) is correct based on reading the passage and looking at the information in the table, which shows that CO_2 is formed when methane is burned. Important information is given under the table, which explains that methane is natural gas. Air pollution (option 1) is not mentioned in the passage. CH_4 (option 3) is methane, which is what is being burned. O_2 (option 4) is what the methane reacts with and is part of the cause. Carbon monoxide (option 5) is not shown on the table.
3. **(1) a decreased population of robins** Option (1) is correct based on understanding that human activity shrinks the robins' habitat, which will impact what they have to eat, the amount of trees in which they can nest, etc., causing a decreased population. Destructive human activity will not

increase woods and meadows (option 2), increase food resources for the robins (option 4), or improve the variety of nesting sites (option 5). Human activity will probably mean fewer robins in the habitat, which will lead to less fertilization (option 3).

4. **(4) Red snappers would move into the estuary.** Option (4) is correct. An increase in salinity farther upstream could allow the red snapper, a saltwater fish, to also swim farther upriver. If saltwater made it farther upstream, gar (option 1) would swim farther upstream, not toward the ocean, because they prefer fresh water. The mussel population would decrease, not increase, due to higher salt levels (option 2). Because of their different tolerances for salinity, mussels would still not share habitat with oysters (option 3) and gar would not share habitat with shrimp (option 5).
5. **(5) crab** Option (5) is correct; the crab has the widest range of tolerance to salinity variations—0.5 to 35 parts per thousand—so the effect of salinity changes would be least among crab populations. Mussels (option 1), freshwater snails (option 2), shrimp (option 3), and starfish (option 4) all have smaller ranges of tolerance for salinity. The variance in salinity would most likely cause those animals to die or move.
6. **(4) passing an electric current through water** Option (4) is correct because the passage states that water breaks down into hydrogen and oxygen gases when an electric current is passed through water. Changing iron (option 1) and producing iron sulfide (option 2) are mentioned in a different process in the passage. There is no mention of mixing sulfur and water (option 3) or of heating solid oxygen and hydrogen (option 5).

Skill 10 Assess Adequacy and Accuracy of Facts
Pages 48–49

1. **(4) Coffee is less acidic than lemon juice.** Option (4) is correct because the figure shows that the pH of coffee is 5 and the pH of lemon juice is 2. Lower numbers are more acidic. Option (1) is inaccurate because the figure includes beverages that are weak acids. Option (2) contradicts the diagram, which shows that coffee is an acid. No facts are presented to support the statement that coffee is more damaging (option 3) or that water is the <u>only</u> neutral substance (option 5).
2. **(2) The pH of blood is between 7 and 10.** Option (2) is correct because the diagram classifies a material with a pH between 7 and 10 as a weak base. Something more basic than coffee (option 1) could still be classified as a weak acid. Having a pH near that of water (option 3) is not specific enough and means that human blood could be a weak acid. Option (4) is not a fact represented in the figure. Having a pH greater than 7 (option 5) is not adequate to classify blood as a weak base because if the pH exceeds 10, it is considered a strong base.
3. **(3) New trees can be planted to replace those that are cut down.** Option (3) is correct. A renewable resource is one that can be replaced, so the ability to plant new trees is adequate to support the conclusion. The fact that there are many trees (option 1) is not adequate because it does not indicate that new trees replace removed trees. Whether trees grow anywhere other than Earth (option 2) does not support the conclusion and is not a part of the passage, so it is not adequate. The statement that renewable resources require billions of years to replace (option 4) is false and contradicts the passage. That trees require billions of years to grow (option 5) is inaccurate and would not support the conclusion that they are a renewable resource.
4. **(1) If we keep using Earth's nonrenewable resources, they will eventually be gone.** Option (1) is correct; the passage supports the idea that nonrenewable resources will run out because they cannot be replaced. The statement that metals and oil are renewable (option 2) contradicts the passage, so it is not accurate. Option (3) may be accurate, but the passage is not adequate to support it because distribution is not mentioned. Options (4) and (5) are incorrect because nothing in the passage supports nonrenewable resources coming from outside Earth or being produced only on Earth.
5. **(2) Taller dandelions would begin to do well in the changed habitat.** Option (2) is correct because the passage indicates that taller dandelions do well in undisturbed fields. There is no data in the passage to support conclusions that few species would grow in the fields (option 1) or that seeds would not be dispersed (option 3), so these statements are not adequately supported. Option (4) is not supported, and option (5) contradicts information from the passage, so they are not accurate based on the information provided.
6. **(3) Armadillos and glyptodonts share a common armadillo-like ancestor.** Option (3) is correct because the diagram shows a common ancestor and the passage indicates it would have similar traits. The diagram indicates that armadillos did not evolve from finches (option 1) and that the recent ancestor is not likely to be bird-like (option 2), so these statements are inaccurate. Finches and armadillos do share a common ancestor (option 4) according to the diagram, so this statement is false. That finches evolved from glyptodonts (option 5) contradicts the information in the diagram and is also false.

KEY Skill 11 Evaluate Information
Pages 52–53

1. **(5) Flowering plants produce seeds after fertilization takes place.** Option (5) is correct because the passage states that seeds start to grow after fertilization. Option (1) is incorrect because male organs produce pollen, not female organs. Anthers (option 2) are part of the male organs, not the female organs. Option (3) is incorrect because pollen descends from the stigma through the pollen tube during pollination. Male sex cells are produced by the stamens (option 4), not the ovaries.
2. **(1) Flowers play an important role in the sexual reproduction of plants.** Option (1) is correct because the flower houses and protects all the key components that allow sexual reproduction to occur. Option (2) is incorrect because neither the passage nor the diagram discusses the ability to pollinate other types of flowers, and in fact, pollen can only fertilize female cells of the same or closely related species. Option (3) is incorrect because the passage implies that the ovary is part of the female sex organs. Option (4) is incorrect because the diagram shows the pollen tube as part of the

female sex organs, not as part of the stamen, which includes the male sex organs. Neither the passage nor the diagram discuss the role of insects in pollination (option 5), and in fact there are many ways for pollination to occur.

3. **(4) Two basic types of root systems are taproots and fibrous roots.** Option (4) is correct because the passage describes the two different root systems in grasses, trees, dandelions, and carrots. The passage does not include information to support leaves absorbing water (option 1), the ability of roots to crack rock (option 2), or the direction of root growth when a pot is placed on its side (option 3). Option (5) is not correct because the passage only talks about water passing in through roots, not passing out of roots.
4. **(3) The amount of displacement of the medium depends on amplitude, not on wavelength.** Option (3) is correct because the diagram shows the displacement as the up and down amplitudes but does not relate displacement to wavelength. Option (1) is incorrect. The passage states that wavelength and frequency are different measures. The energy of a wave (option 2) and information about light and sound waves (option 5) are not presented in the passage or diagram, so these choices are incorrect. Option (4) cannot be inferred from the passage.
5. **(4) amplitude** Option (4) is correct because the diagram clearly shows the amplitude as the maximum displacement or distance above or below the rest position as shown by the arrows. The trough (option 1) and crest (option 2) are the highest and lowest points of the wave, clearly labeled on the diagram. The wavelength (option 3) and frequency (option 5) are measurements of the wave, not related to its displacement.
6. **(1) The distance from one trough to the next is one wavelength.** Option (1) is correct because the diagram shows that the distance between crests is the same as the distance between troughs—one wavelength. Option (2) is incorrect because waves can travel from any one place to another, in any direction. Option (3) is incorrect because wavelength and amplitude are independent and measure different wave properties. Option (4) is incorrect because the diagram shows that particles move up and down, not in the same direction as the wave. Although the diagram shows 4 crests and 3 troughs (option 5), evaluation of the concepts shown reveals that the ratio of crests to troughs is 1:1.

KEY Skill 12 Recognize Faulty Logic
Pages 56–57

1. **(3) Brown alleles are most likely dominant because most people have brown eyes.** Option (3) is correct because anyone with the dominant brown allele shows its phenotype; therefore, it will be more common. There is nothing to indicate eye color is determined by a single gene because both eyes are the same color (option 1). There are dominant and recessive traits for every gene, so the number of genes involved does not indicate dominance (option 2). Option (4) relies on the faulty assumption that the information in the graph applies to all populations. There is no indication that darkness is correlated to trait dominance (option 5).
2. **(1) The phenotype of a trait is determined by the dominant gene.** Option (1) is correct because the passage states that the dominant gene determines the phenotype. The other answers are built on faulty logic. Option (2), that an organism has only one allele, contradicts the passage. Option (3) is incorrect; dominance of a trait does not indicate that the trait is better. Option (4) is incorrect; some people will not have the allele for brown eyes. Every gene has alleles, not just genes for eye color (option 5).
3. **(4) The motion of an object is affected by all the forces that act on it.** Option (4) is correct because the passage states that motion is the result of forces. Option (1) is incorrect. Even moving objects that do not stop are affected by friction. Option (2) is incorrect; friction affects all surfaces in contact with one another. Option (3) is incorrect. There is a gravitational force between any two objects, as stated in the passage. Option (5) is inaccurate and not supported by information in the passage.
4. **(3) Lightning does not occur between two objects with the same electrical charge.** Option (3) is correct because lightning is caused by the attraction between opposite charges. Option (1) is incorrect; because it rains often when lightning occurs doesn't mean rain is necessary to produce lightning. The passage states that *some* of the energy is converted to light and sound (option 2), but that doesn't mean *most* of the energy is released this way. Option (4) is incorrect because there is no indication that the cloud must be stationary to form charged regions. Lightning is dangerous because it is a powerful electric current (option 5), and it is faulty logic to assume that electric current is not dangerous.
5. **(5) Lightning cannot strike the tree because there is a positive charge on the tree.** Option (5) is correct because this contradicts the passage, which states that electrons flow to a positively charged region; this means that lightning is more likely to strike a tree. It is logical to state that lightning can travel within a cloud (option 1) because the diagram shows regions of positive and negative charge in the cloud. Option (2) is incorrect because the passage states that lightning occurs when electrons travel between areas of opposite charge. The positive charge on the tree is closer to the cloud, so it is a more likely place to strike (option 3). Option (4) is incorrect because the passage states that the energy comes from the flow of charged particles.
6. **(3) Nuclear-powered subs can remain submerged for a much longer time.** Option (3) is correct because a submarine under the ice must remain submerged for a long period. Options (1) and (2) describe attributes of the nuclear subs but no logical support for the conclusion in question. The fact the United States had nuclear subs first (option 4) does not logically support the conclusion. Option (5) is incorrect; the reliance on batteries is a logical argument against diesel subs but not for nuclear subs.

KEY Skill 13 Tables and Charts
Pages 60–61

1. **(4) amphibians** Option (4) is correct because amphibians and mammals are part of the same kingdom. Option (1) is not correct because bacteria are not listed in the chart. Protozoans

Protozoans (option 2) are classified as Protista (row 2), and mammals are Animalia (row 5). Plantae (option 3) is the name of a kingdom that does not include mammals. Fungi (option 5) are in the Fungi kingdom, so they are in different kingdoms, and therefore less like mammals than are amphibians.

2. **(3) by photosynthesizing food** Option (3) is correct because the table indicates that these organisms are all part of the kingdom Plantae, which only photosynthesizes food. Absorbing food from the environment (options 1 and 5) is shown for Protista and Fungi only. Ingesting nutrients (options 2, 4, and 5) is shown in the table only for Protista and Animalia.
3. **(2) pancreas** Option (2) is correct. According to the table, the pancreas controls insulin production, and diabetes is caused by the insufficient production of insulin. Options (1), (3), (4), and (5) are incorrect because they control immunity, the nervous system, weight, and salt/carbohydrate metabolism, respectively, none of which are connected to diabetes or insulin.
4. **(1) Everyone in Santa Cruz feels the ground shake, but no major damage occurs.** Option (1) is correct because the table indicates that a 4.9 to 5.4 earthquake would be felt by everyone, but there is no indication that it would cause damage. The earthquakes that are not felt by everyone in the area (options 2 and 3) are classified lower than 4.9 on the Richter Scale. Earthquakes that cause damage (options 4 and 5) are classified higher than 5.4 on the Richter Scale.
5. **(2) A wrecking ball flattens a row of old houses.** Option (2) is correct because the table describes an earthquake in the 7.4 to 7.9 range as causing buildings to collapse. A truck (option 1) or train (option 3) causing the ground to shake, a falling cabinet causing minor damage (option 4), and broken windows in a storm (option 5) would all correspond to less severe earthquakes with a magnitude lower than 7.4 on the Richter Scale.
6. **(3) comparing the strength of earthquakes in Tokyo and Calcutta** Option (3) is correct because one can infer from the information in the table that the purpose of the scale is to rank earthquakes by their effects, which allows them to be compared. There is no indication on the table of the number of earthquakes in a particular location (option 1), the actual number of buildings damaged (option 2), or the speed of earthquake waves (option 5). Seismographs are instruments used to measure the magnitude of an earthquake according to the Richter Scale, so option (4) is incorrect.

KEY Skill 14 Bar Graphs

Pages 64–65

1. **(1) apple** Option (1) is correct. According to the bar graph, an apple has roughly 75 calories, the same amount the average 30-year-old woman burns during a 1-mile walk. Options (2), (3), (4), and (5) all have more than 75 calories.
2. **(2) tomato** Option (2) is correct because the graph shows about 40 calories per 200-gram tomato serving. Beef steak (option 1) has 280 calories per serving; rice (option 3) has 225 calories per serving; potatoes (option 4) have almost 150 calories per serving; and bananas (option 5) have 100 calories per serving. These options have values that are all greater than tomatoes.
3. **(2) spring (Apr–Jun)** The best time for scientists to study the formation of tornadoes is the time period with the greatest number of tornadoes. Option (2) is correct because the total number of tornadoes during the 3-month spring period is about 840, which is the greatest number. The other choices—winter (option 1) has about 80 tornadoes, summer (option 3) has about 430, and autumn (option 4) has about 160 tornadoes—all have fewer than 840 tornadoes. Option (5) contradicts the information on the graph because the varying bar lengths indicate that the most tornadoes occur in spring.
4. **(4) 850** Option (4) is correct because there are about 400 tornadoes in May, about 300 in June, and about 150 in July, for an estimated total of 850. An estimate of 280 (option 1) would be about the average per month during the period. About 400 tornadoes (option 2) is the value for May alone. An estimate of 700 (option 3) only reflects May and June–it does not include July. A value of 1,500 (option 5) is an estimate of the total number of tornadoes for all months.
5. **(2) Volume increased in all three liquids from 40°C to 60°C.** According to the graph, each of the three liquids showed an increase in volume as the heat increased from 40°C to 60°C—petroleum increased about 2.5%; chloroform increased nearly 3%; and water increased about 0. 25%—so option (2) is correct. Option (1) is incorrect because the volume of the liquids changed by varying amounts, not the same amount. Option (3) is not correct because water showed an increase, not decrease, with increased heat. Option (4) is incorrect because the percentage change in the volumes of the three liquids was measurable. There is not enough information to determine the cost of chloroform or petroleum, so option (5) is incorrect.
6. **(1) Expansion by heat is dependent on the kind of liquid.** Option (1) is correct because according to the graph, the volume of each liquid increased by different amounts, which indicates that expansion relies, or depends, on the kind of liquid. Option (2) is incorrect because there is no information about gases in the graph. Option (3) is incorrect because the graph shows the volume increases as the temperature increases, which means volume is dependent on temperature, not independent. Option (4) is incorrect because petroleum and chloroform increased by different amounts. Option (5) is not correct because there is no information about liquids at 80°C or the explosiveness of liquids.
7. **(4) increase by an unknown amount** All the bars on the graph show an increase from 40°C to 60°C, so it can be inferred that the liquid's volume will increase when heated. It is not possible, however, to determine how much the volume will increase without knowing the liquid's identity, so option (4) is correct. Based on the information in the graph, the volume of the unknown liquid will increase, so option (1) (no change) and option (5) (decrease) are not correct. Options (2) and (3) are not correct because without knowing what the liquid is, it is not possible to determine the amount of increase.
8. **(4) water** Option (4) is correct. Based on the bar graph, you can see that water shows the least change in volume when heated. From this you can conclude that water is the least affected by changes in temperature of the three liquids. Option (1) is incorrect because the graph shows the liquids reacting differently to changes in temperature. Options (2) and (3)

are incorrect because both petroleum and chloroform were affected more by temperature changes than water. Option (5) is incorrect because water was clearly the least affected of the three.

KEY Skill 15 Line Graphs

Pages 68–69

1. **(1) forest** Option (1) is correct because the graph shows that shade plants have a higher rate of photosynthesis than sun plants when there is low light intensity, such as in the shade of a forest. The desert (option 2), a sunny meadow (option 3), and a sunny windowsill (option 4) are places where there is high light intensity, so sun plants will have a higher rate of photosynthesis. A dark cave (option 5) has no light, so photosynthesis would not occur there.
2. **(2) 375** Option (2) is correct because the two lines cross about ¾ of the way between the 200 and 400 lines of the graph, or about 375 photons/m^2/s. At the value 200 (option 1), the blue line has a greater value, so shade plants have a higher rate. At the values 425, 500, and 650 (options 3, 4, and 5), the sun plants already have a clearly greater rate.
3. **(2) 1.8** Option (2) is correct because the line indicating a ban on CFCs crosses the 2050 line at a level slightly less than 2 parts per million (ppm). The lowest value (option 1) could be a result of reading the scale lines as 1 ppm, not 2 ppm. The value 2.2 ppm (option 3) would be slightly above the 2 ppm line, not below. The value 6.4 ppm (option 4) would be correct if CFC use is reduced. The value 11.8 ppm (option 5) would be correct if there are no CFC restrictions.
4. **(5) The average speed is not the same as the actual speed.** Option (5) is correct because the graph shows the train traveling 60 miles per hour at 0 hours, which cannot be true because the train would have been stopped at the beginning of the trip (0 hours). It is assumed the reader knows the word *average* is a calculation based on the actual speed. Option (1) is not correct because the average speed remains the same as the distance increases. Options (2) and (3) are incorrect because the average speed is shown as a steady constant, not as constantly increasing or decreasing. Option (4) is wrong; only the average speed is given—the actual speed probably did vary to some degree during the trip.
5. **(4) 180** Option (4) is correct because according to the distance-time graph, the train had traveled 180 miles at 3 hours. The other options are based on a misunderstanding of the graphs. The number 4 (option 1) is shown as a measure of time on both graphs. The number 60 (option 2) is the average speed of the train in miles per hour. The number 120 (option 3) is the distance traveled after only 2 hours. The number 240 (option 5) is the distance traveled after 4 hours.
6. **(4) The graphs show two aspects of the same trip.** Option (4) is correct because one graph shows the relationship between distance and time during a four-hour period, while the other graph shows the relationship between average speed and time during a four-hour period. Options (1), (3), and (5) are not correct because they contradict the information shown on the axes of the graphs. Option (2) is incorrect because according to the graphs, distance constantly increased as the average speed remained the same, so they cannot be the same measure.
7. **(3) As time increases, distance also increases.** Option (3) is true because according to the graphs, the distance increased by 60 miles during each hour of the trip. Options (1) and (2) are not correct because the average speed remained constant; it did not increase. Options (4) and (5) are not correct because time cannot decrease or remain the same; it is always increasing.

KEY Skill 16 Circle Graphs

Pages 72–73

1. **(2) listeria** Option (2) is correct because according to the graph, listeria caused 0.2 million food-related illnesses, which is the smallest number on the graph. All the other options represent a number larger than listeria: botulism – 0.58 million (option 1), campylobacter – 1 million (option 3), salmonella – 2.3 million (option 4), and Norwalk virus – 9.2 million (option 5).
2. **(4) Norwalk virus** Option (4) is correct because the area of the circle corresponding to the Norwalk virus is clearly greater than half the circle. Campylobacter and salmonella combined (option 1); botulism and salmonella combined (option 2); listeria, campylobacter, and other microbes combined (option 4); and salmonella alone (option 5) all occupy an area less than one-quarter of the circle, so they cannot be the correct answer.
3. **(1) Silicon and oxygen are the two most abundant elements of Earth's crust.** Silicate is composed of silicon and oxygen, and according to the graph, silicon and oxygen are the most abundant elements in Earth's crust, so it can be concluded that the abundance of silicon and oxygen in Earth's crust correlates directly to the silicate composition of rocks on Earth's surface. Option (2) is incorrect because no information is provided about the hardness of iron. Option (3) is incorrect because no information is provided about the ability of elements to combine with other elements. Option (4) is incorrect because it contradicts information in the graph, which shows that iron, aluminum, and trace elements are also elements in Earth's crust. The graph does not provide information about the depth of the elements, so option (5) is not correct.
4. **(4) increase in toxin-producing algae** According to the graph titled *Causes of UMEs*, biotoxins represent 27% and are the second largest cause of UMEs. Since no answer choice corresponds to the largest section, unknown causes, and since toxin-producing algae are classified as biotoxins, option (4) is correct. All the other answer options are less than 27%. Option (1) is incorrect because an oil spill would be classified as human interaction, which is only 6% of the causes. A decline in seaweed (option 2) would be an ecological factor that causes UMEs, but ecological factors do not represent a large portion of the causes. Option (3) is incorrect because a new virus would be classified as an infectious disease, and this only represents 13% of the graph. The decreased use of fishing boats (option 5) would most likely cause a decrease in UMEs, so it is not the correct choice.
5. **(4) human interactions** Option (4) is correct because the section for human interactions is the smallest section of the graph at only 6% and therefore represents the least likely cause. Infectious diseases (option 1), biotoxins (option 2),

and ecological factors (option 3) are all indicated with larger sections of the circle than human interactions. The graph sections are clearly not equal in size, so option (5) is not correct.

6. **(4) Manatees are less affected by UMEs than cetaceans.** Option (4) is correct because according to the graph, the cetaceans are the group most affected by the UMEs, which means that all other groups, including manatees, are less affected. For this same reason, options (3) and (5) are incorrect – cetaceans are more affected than manatees and pinnipeds. Option (1) is incorrect because while the cetaceans have the largest percent of population affected by UMEs, there is no information about actual population size. Option (2) is incorrect because there is no information to indicate that manatees swim faster than sea otters.
7. **(2) sea otters** Option (2) is correct because the segment of the diagram corresponding to sea otters represents the smallest portion of the circle. Cetaceans (option 1), pinnipeds (option 3), manatees (option 4), and the combination of manatees and sea otters (option 5) are all indicated by sections that cover larger portions of the circle.

KEY Skill 17 Diagrams Pages 76–77

1. **(1) shale above the sandstone** Option (1) is correct because the problem states that the cap rock is impermeable and above the oil, and the diagram indicates that shale is impermeable. Sandstone (option 2) is a permeable rock, so it cannot be the cap rock. The limestone does not extend all the way across the top of the oil (option 3). The shale (option 4) and limestone (option 5) that are beneath the sandstone cannot be cap rock because the oil is above them.
2. **(1) When drilling you are likely to hit gas first, then oil, and then water.** Option (1) is correct because the diagram shows that the gas layer is closest to the surface, then the oil, and then water. The diagram indicates that oil collects within a sandstone rock layer, not as an underground lake (option 2). Option (3) contradicts the diagram because gas-filled sandstone is shown above the oil, not below. Shale is labeled as impermeable, so liquids can't flow through it (option 4). There is no information about water wells (option 5).
3. **(5) Isotopes have the same number of protons but different numbers of neutrons.** Option (5) is correct because the diagram clearly shows three different isotopes that differ only by the number of neutrons. Hydrogen is an element (option 1), but that information is not shown by the diagram. The diagram shows that deuterium is a form of hydrogen (option 2). Option (3) contradicts the diagram because each hydrogen isotope has a different number of neutrons. There is no indication that other elements do not have isotopes (option 4) because the diagram focuses only on hydrogen.
4. **(1) The lungs breathe in air and oxygen through the nose and mouth.** Option (1) is correct because the diagram indicates that both types of respiration form a cycle that starts with an arrow indicating an intake of air and oxygen. The transfer of oxygen by blood (option 2), production of carbon dioxide by cells (option 3), and transfer of carbon dioxide to the lungs (option 4) are indicated as steps of internal respiration, not external. The exchange of gas (option 5) does not occur until after the beginning of external respiration.
5. **(3) gills, which exchange carbon dioxide for oxygen in water** Option (3) is correct because the diagram indicates that the function of the lungs is to exchange gases. The lateral line (option 1), scales (option 2), and kidneys (option 4) perform functions unrelated to respiration. There is no parallel to these organs in the diagram. The heart (option 5) pumps blood through the body, which is a different internal function than that of the lungs.
6. **(2) an increase in the rate at which air and oxygen are inhaled and exhaled** Option (2) is correct because oxygen is used and carbon dioxide is produced during cellular respiration, so an increased level of cellular respiration would increase the rate of breathing. A decrease in carbon dioxide production (option 1) and a decrease in the heart rate (option 3) would be caused by a decrease in respiration, not an increase. According to the diagram, oxygen is not produced by cellular respiration (option 4). Because cellular respiration produces energy, an increased level would increase energy production (option 5), not decrease it.
7. **(3) energy** Option (3) is correct because the diagram indicates that cellular respiration produces energy, and an increased heart and lung capacity would increase the level of cellular respiration, therefore increasing energy. The amount of muscle mass (option 1), strength (option 2), coordination (option 4), and balance (option 5) are not mentioned in the diagram in relation to respiration.

KEY Skill 18 Combine Text and Graphics Pages 80–81

1. **(4) the western coast** Option (4) is correct because the map shows that the western coast of the United States is along a plate boundary, and the passage states that volcanoes occur more frequently along plate boundaries. The northeast coast (option 1), southeast coast (option 2), and central region (option 3) are not near plate boundaries according to the map. Option (5) is incorrect because the west coast has a plate boundary and the others do not.
2. **(3) continue to occur along plate boundaries** Option (3) is correct because the passage states that 80% of Earth's surface was caused by volcanoes, and that 500 volcanoes are still active, indicating that they are a continual process. The map and passage show that they still occur along boundaries. There is no indication that volcanic activity will cease (option 1). They are likely to continue along the Ring of Fire, not decrease (option 2), due to the location of plate boundaries. There is no indication in the passage or map about the destruction of major cities (option 4) or the effect of global warming on volcanic activity (option 5).
3. **(3) New crust is generated on the ocean floor by volcanoes along the line.** Option (3) is correct because the passage states that new ocean floors are caused by volcanoes, and according to the map, the line is a divergent plate boundary, and new crust is generated along divergent plate boundaries. The information in the passage indicates that the Ring of Fire is in the Pacific Ocean (option 1), and volcanoes occur in places other than the Ring of Fire (option 2). There is no indication in the passage or on the map that new continents will form (option 4). Option (5) contradicts

information on the map about the definition of divergent plate boundaries—crust is consumed along convergent plate boundaries.

4. **(2) There are few plate boundaries within the continents.** Option (2) is correct because the passage indicates that volcanoes are most likely at plate boundaries, and the map shows that there are very few plate boundaries within continents. There is no information to indicate that the weight of the oceans causes volcanoes (option 1) or that the crust is thicker in the middle of a continent (option 4). Only about half of volcanoes occur along the Ring of Fire (option 3). Volcanoes occur along both types of plate boundaries (divergent and convergent) and at other places (option 5).
5. **(1) Every radioisotope has its own constant rate of decay and half-life.** Option (1) is correct because the characteristic half-lives of elements in the rock can be used, along with the remaining amount of the material, to determine how long the isotope has been in the rock. The half-life of thorium-234 is too short for its use in dating rocks (options 2 and 4). The statements about radioisotopes (option 3) and half-life (option 5) are definitions, not support for the conclusion.
6. **(4) On the 24th day, only 8 of the original 16 grams of thorium-234 remain.** Option (4) is correct because the half-life is defined as the period of time required for half the material to decay. After 2 days, 8 grams of thorium-234 have decayed, not 4 (option 1). After 48 days, 12 grams of thorium-234 have decayed, not 8 (option 2). While it is true that every radioisotope has a characteristic half-life (option 3) and that some half-lives are less than one second (option 5), those facts do not support the statement or the graph.
7. **(5) 2** Option (5) is correct because the passage indicates that the amount will be halved 3 times, and the graph shows 2 grams remaining after 72 days. The other answer choices are amounts that will remain after periods less than 3 half-lives: 16 grams (option 1) is the original sample, 8 grams (option 2) is the amount remaining after 1 half-life, 6 grams (option 3) is between 1 and 2 half-lives, and 4 grams (option 4) is 2 half-lives.
8. **(2) 1:1** Option (2) is correct because half of the uranium-238 will have decayed after one half-life, so there will be equal amounts of uranium-238 and lead-205. The ratio 1:0 (option 1) indicates that no decay will have occurred. A 1:2 ratio (option 3) and a 1:4 ratio (option 5) both indicate that more than half of the material has decayed, which will occur in more than 4.5 billion years. A 2:1 ratio (option 4) occurs in a period of less than one half-life.

KEY Skill 19 Combine Information from Graphics
Pages 84–85

1. **(5) what a negative charge is** Option (5) is correct because based on the the diagrams, the reader can assume that a negative charge is a property of an electron. Option (1) contradicts the first diagram. The diagrams show that the neutral object has no extra electrons (option 2) and how uncharged objects get negative charges (option 3), so it is not assumed that the reader knows this. Positive charges (option 4) are not related to the diagram, so no knowledge of them is assumed.
2. **(2) to demonstrate that electrons can be transferred between objects that touch one another** Option (2) is correct because the diagrams show an uncharged object first, then show the object touching a charged object, and then show the objects repelling. There is no indication in the diagrams that all objects contain electrons (option 1) or that there is a danger of electric shock (option 5). The color change (option 3) is just a property that the artist used to show a difference between charged and uncharged objects. The subject of the series is the transfer of electrons, not the properties of balloons (option 4).
3. **(5) The balls will move together, touch, and then move apart.** Option (5) is correct because the text in the diagrams indicates that it applies to charged objects, not just balloons, so table tennis balls should show the same effects. The statement that the balls will move apart (option 1) will happen, but other steps need to occur first. Movement (option 2) is shown in all of the diagrams, so the balls will most likely move. Positive charges (option 3) are not discussed in the diagrams. The diagrams do not show a negatively charged particle losing its charge (option 4), only objects gaining charge.
4. **(3) two negatively charged objects** Option (3) is correct because the final diagram indicates that both objects have a negative charge. The other choices, two positively charged objects (option 1), two neutral objects (option 2), one negative and one positive (option 4), and one negative and one neutral (option 5), all contradict information in the third diagram.
5. **(3) 400** Option (3) is correct. You can see from the table that a simple pulley doubles the input force. So if the man exerts an input force of 200, the output force, or the force that moves the box, will be doubled to 400. Option (1) is incorrect and may be the result of reading the table incorrectly. Option (2) is incorrect. A simple pulley doubles the input force; the output force would not be the same as the input force. Options (4) and (5) are both incorrect and more than the correct output force.
6. **(4) stapes** Option (4) is correct because the Ear Anatomy diagram shows that contact with the cochlea occurs at the structure labeled "stapes" in the Middle Ear diagram. The eardrum (option 1), incus (option 2), and malleus (option 3) are all parts of the inner ear that transmit vibrations, but they are not in direct contact with the cochlea, according to the Ear Anatomy diagram. The vestibular cochlear nerve (option 5) is not part of the middle ear.
7. **(3) a middle ear infection** Option (3) is correct because the middle ear is attached to the inner ear, so an infection in the middle ear could cause damage to the inner ear, resulting in conductive hearing loss. The auditory nerve (option 2) is not addressed and therefore cannot be the correct answer choice. Since conductive hearing loss is caused by damage to the inner ear, injury to the outer ear (option 5) and buildup in the auditory canal (option 1), which is a part of the outer ear, are not likely to cause damage to the inner ear. According to the diagram, the eustacian tube (option 4) connects the middle ear to other parts of the body, not the inner ear.

KEY Skill 20 Use the Scientific Method

Pages 88–89

1. **(1) Carbon dioxide and oxygen are exchanged in the lungs.** Option (1) is correct because the diagram shows that the exchange of carbon dioxide and oxygen occurs in the alveoli, which are part of the lungs. While people with bronchitis do tend to cough (option 2), this is not a conclusion based on the passage or diagram. The passage discusses a network of tiny vessels, not large, so option (3) is incorrect. Option (4) contradicts the diagram, which gives no information about the larynx. Option (5) is incorrect because the size of oxygen and carbon dioxide molecules is not shown or discussed.
2. **(3) The smoke damages the alveoli.** Option (3) is correct because the diagram shows that the alveoli are responsible for the exchange of oxygen and carbon dioxide. If that exchange system does not function well, then oxygen is not absorbed effectively. While smoke may affect the pharynx (option 1), that is not indicated to affect oxygen supply. According to the passage, carbon dioxide (option 2) is a waste material, so there is no effectiveness to decrease. Whether smokers exercise less (option 4) is not observed or discussed. The data provided does not indicate that smokers have a greater number of alveoli than nonsmokers (option 5).
3. **(1) reduced transfer of oxygen to the capillaries** Option (1) is correct because the observations suggest that bronchioles carry air to the alveoli, so liquid accumulation in the bronchioles would probably reduce their effectiveness. For the same reason, the liquid would reduce the removal of carbon dioxide, not make it more efficient (option 2). The bronchioles are not observed to be near the nose and pharynx, so they would be unlikely to cause irritation there (option 3). Decreased transfer of gases would likely cause an increased heart rate, not a slower one (option 4), as the body tries to remove carbon dioxide. There is no indication that growth of new alveoli (option 5) can occur.
4. **(2) Use tags to identify birds from several places and find out where they migrate.** Option (2) is correct because this experiment would allow scientists to study a representative sample of the population. Following each bird (option 1) would be nearly impossible so it is not a reasonable experiment. The lengths of birds (option 3) can vary so that is not a reliable way to identify populations. Counting the birds (option 4) will not indicate their point of origin. Feeding habits are more likely to be influenced by available food than by migration routes (option 5).
5. **(1) lack of food in cold winter areas** Option (1) is correct because the food supply in an area is limited, so animals migrate to areas with larger food supplies. The passage also states that migratory patterns are probably related to the animals' need for food. It is easier to adapt to cold (option 2) and to change breeding patterns (option 3) than to adapt to a food shortage. Increased competition in the tropics (option 4) would discourage migration, not encourage it. There is no indication that natural enemies increase in cold winter areas (option 5).
6. **(4) Short-tailed shearwaters will begin their migration in March or April.** Option (4) is correct because the passage states that food availability is a primary reason for migration, and food would become less plentiful in the southern hemisphere in March or April as fall begins. Arctic terns are likely to move toward the north in March and toward the south in September (option 1) because those periods mark the beginning of spring. Monarchs do not stay in one location (option 2). The monarchs would fly south in the Northern Hemisphere during the fall, not spring (option 3). The map clearly shows several separate migrations of blue whales (option 5).

Pretest Answer Sheet: Science

Name: ______________________ Class: __________ Date: __________

1 ①②③④⑤	5 ①②③④⑤	9 ①②③④⑤	13 ①②③④⑤	17 ①②③④⑤
2 ①②③④⑤	6 ①②③④⑤	10 ①②③④⑤	14 ①②③④⑤	18 ①②③④⑤
3 ①②③④⑤	7 ①②③④⑤	11 ①②③④⑤	15 ①②③④⑤	19 ①②③④⑤
4 ①②③④⑤	8 ①②③④⑤	12 ①②③④⑤	16 ①②③④⑤	20 ①②③④⑤

Official GED Practice Test Form PA Answer Sheet: Science

Name: ______________________ Class: __________ Date: __________

Time Started: __________

Time Finished: __________

1 ①②③④⑤	6 ①②③④⑤	11 ①②③④⑤	16 ①②③④⑤	21 ①②③④⑤
2 ①②③④⑤	7 ①②③④⑤	12 ①②③④⑤	17 ①②③④⑤	22 ①②③④⑤
3 ①②③④⑤	8 ①②③④⑤	13 ①②③④⑤	18 ①②③④⑤	23 ①②③④⑤
4 ①②③④⑤	9 ①②③④⑤	14 ①②③④⑤	19 ①②③④⑤	24 ①②③④⑤
5 ①②③④⑤	10 ①②③④⑤	15 ①②③④⑤	20 ①②③④⑤	25 ①②③④⑤